ENVIRONMENTAL SCIENCE, ENGINEERING AND TECHNOLOGY

REDUCING GREENHOUSE GAS EMISSIONS

AVAILABLE AND EMERGING TECHNOLOGIES

ENVIRONMENTAL SCIENCE, ENGINEERING AND TECHNOLOGY

Additional books in this series can be found on Nova's website under the Series tab.

Additional E-books in this series can be found on Nova's website under the E-books tab.

ENVIRONMENTAL SCIENCE, ENGINEERING AND TECHNOLOGY

REDUCING GREENHOUSE GAS EMISSIONS

AVAILABLE AND EMERGING TECHNOLOGIES

DIANE B. MCCREEVEY

AND

ELLEN L. DURKIN

EDITORS

Nova Science Publishers, Inc.

New York

For permission to use material from this book please contact us:
Telephone 631-231-7269; Fax 631-231-8175
Web Site: http://www.novapublishers.com

NOTICE TO THE READER

The Publisher has taken reasonable care in the preparation of this book, but makes no expressed or implied warranty of any kind and assumes no responsibility for any errors or omissions. No liability is assumed for incidental or consequential damages in connection with or arising out of information contained in this book. The Publisher shall not be liable for any special, consequential, or exemplary damages resulting, in whole or in part, from the readers' use of, or reliance upon, this material. Any parts of this book based on government reports are so indicated and copyright is claimed for those parts to the extent applicable to compilations of such works.

Independent verification should be sought for any data, advice or recommendations contained in this book. In addition, no responsibility is assumed by the publisher for any injury and/or damage to persons or property arising from any methods, products, instructions, ideas or otherwise contained in this publication.

This publication is designed to provide accurate and authoritative information with regard to the subject matter covered herein. It is sold with the clear understanding that the Publisher is not engaged in rendering legal or any other professional services. If legal or any other expert assistance is required, the services of a competent person should be sought. FROM A DECLARATION OF PARTICIPANTS JOINTLY ADOPTED BY A COMMITTEE OF THE AMERICAN BAR ASSOCIATION AND A COMMITTEE OF PUBLISHERS.

Additional color graphics may be available in the e-book version of this book.

Library of Congress Cataloging-in-Publication Data

ISBN 978-1-61470-726-4

Published by Nova Science Publishers, Inc. † New York

CONTENTS

PREFACE

This book examines control techniques and measures to mitigate greenhouse gas (GHG) emissions from specific industrial sectors such as coal-fired electric generating units; the petroleum industry and the iron and steel industry.

Chapter 1- This document is one of several white papers that summarize readily available information on control techniques and measures to mitigate greenhouse gas (GHG) emissions from specific industrial sectors. These white papers are solely intended to provide basic information on GHG control technologies and reduction measures in order to assist States and local air pollution control agencies, tribal authorities, and regulated entities in implementing technologies or measures to reduce GHGs under the Clean Air Act, particularly in permitting under the prevention of significant deterioration (PSD) program and the assessment of best available control technology (BACT) These white papers do not set policy, standards or otherwise establish any binding requirements; such requirements are contained in the applicable EPA regulations and approved state implementation plans.

Chapter 2- This document is one of several white papers that summarize readily available information on control techniques and measures to mitigate greenhouse gas (GHG) emissions from specific industrial sectors. These white papers are solely intended to provide basic information on GHG control technologies and reduction measures in order to assist States and local air pollution control agencies, tribal authorities, and regulated entities in implementing technologies or measures to reduce GHGs under the Clean Air Act, particularly in permitting under the prevention of significant deterioration (PSD) program and the assessment of best available control technology (BACT). These white papers do not set policy, standards or otherwise establish any binding requirements; such requirements are contained in the applicable EPA regulations and approved state implementation plans.

Chapter 3- This document is one of several white papers that summarize readily available information on control techniques and measures to mitigate greenhouse gas (GHG) emissions from specific industrial sectors. These white papers are solely intended to provide basic information on GHG control technologies and reduction measures in order to assist States and local air pollution control agencies, tribal authorities, and regulated entities in implementing technologies or measures to reduce GHGs under the Clean Air Act, particularly in permitting under the prevention of significant deterioration (PSD) program and the assessment of best available control technology (BACT). These white papers do not set policy, standards or otherwise establish any binding requirements; such requirements are contained in the applicable EPA regulations and approved state implementation plans.

In: Reducing Greenhouse Gas Emissions
Editors: Diane B. McCreevey and Ellen L. Durkin

ISBN: 978-1-61470-726-4
© 2011 Nova Science Publishers, Inc.

Chapter 1

AVAILABLE AND EMERGING TECHNOLOGIES FOR REDUCING GREENHOUSE GAS EMISSIONS FROM COAL-FIRED ELECTRIC GENERATING UNITS[*]

United States Environmental Protection Agency

ACRONYMS AND ABBREVIATIONS

APFBC	Advanced pressurized fluidized bed combustion
ASTM	American Society for Testing and Materials
ASME	American Society of Mechanical Engineers
ASU	Air separation unit
BACT	Best Available Control Technology
Btu	British thermal unit
CAA	Clean Air Act
CCS	Carbon capture and storage
CEMS	Continuous emission monitoring system
CFB	Circulating fluidized bed
CH4	Methane
CO	Carbon monoxide
CO$_2$	Carbon dioxide
EGU	Electric generating unit
EPA	U.S. Environmental Protection Agency
FBC	Fluidized bed combustion
EPRI	Electric Power Research Institute
FGD	Flue gas desulfurization
GHG	Greenhouse gas
H2O	Water

---

[*] This is an edited, reformatted and augmented version of the United States Environmental Protection Agency publication, dated October 2010.

HRSG	Heat recovery steam generator
HHV	Higher heating value
IGCC	Integrated gasification combined cycle
IEA	International Energy Agency
kJ	Kilojoule
kW	Kilowatt
kWh	Kilowatt-hour
LCOE	Levelized cost of electricity
Mg	Megagram
MMBtu/hr	Million Btu per hour
MPa	Megapascal
MW	Megawatt
MWe	Megawatt electrical
MWh	Megawatt-hour
MSW	Municipal solid waste
N2O	Nitrous oxide
NETL	National Energy Technology Laboratory
NOX	Nitrogen oxides
O&M	Operation and maintenance
PC	Pulverized coal
PFBC	Pressurized fluidized bed combustion
PM	Particulate matter
PRB	Power River Basin
scfm	Standard cubic feet per minute
SO_2	Sulfur dioxide
SO_3	Sulfur trioxide
SNCR	Selective noncatalytic reduction
ton/day	tons per day
ton/yr	tons per year
U.S. DOE	U.S. Department of Energy
U.S. EIA	U.S. Energy Information Administration

1. INTRODUCTION

This document is one of several white papers that summarize readily available information on control techniques and measures to mitigate greenhouse gas (GHG) emissions from specific industrial sectors. These white papers are solely intended to provide basic information on GHG control technologies and reduction measures in order to assist States and local air pollution control agencies, tribal authorities, and regulated entities in implementing technologies or measures to reduce GHGs under the Clean Air Act, particularly in permitting under the prevention of significant deterioration (PSD) program and the assessment of best available control technology (BACT). These white papers do not set policy, standards or otherwise establish any binding requirements; such requirements are contained in the applicable EPA regulations and approved state implementation plans.

This document provides information on control techniques and measures that are available to mitigate GHG emissions from the coal-fired electric generating sector at this time. The primary GHG emitted by the coal-fired electric generation industry is carbon dioxide (CO2), and the control technologies and measures presented in this document focus on this pollutant. While a large number of available technologies are discussed here, this paper does not necessarily represent all potentially available technologies or measures that that may be considered for any given source for the purposes of reducing its GHG emissions. For example, controls that are applied to other industrial source categories with exhaust streams similar to the cement manufacturing sector may be available through "technology transfer" or new technologies may be developed for use in this sector.

The information presented in this document does not represent U.S. EPA endorsement of any particular control strategy. As such, it should not be construed as EPA approval of a particular control technology or measure, or of the emissions reductions that could be achieved by a particular unit or source under review.

1.1. Electric Power Generation Using Coal

Electricity is generated at most electric power plants by using mechanical energy to rotate the shaft of electromechanical generators. The mechanical energy needed to rotate the generator shaft can be produced from the conversion of chemical energy by burning fuels or from nuclear fission; from the conversion of kinetic energy from flowing water, wind, or tides; or from the conversion of thermal energy from geothermal wells or concentrated solar energy. Electricity also can be produced directly from sunlight using photovoltaic cells or by using a fuel cell to electrochemically convert chemical energy into an electric current.

In 2008, approximately 70% of the electricity used in the United States was generated by burning fossil fuels (coal, natural gas, petroleum liquids) (U.S. EIA 2010). The combustion of a fossil fuel to generate electricity can be either: 1) in a steam generating unit (also referred to simply as a "boiler") to feed a steam turbine that, in turn, spins an electric generator: or 2) in a combustion turbine or a reciprocating internal combustion engine that directly drives the generator. Some modern power plants use a "combined cycle" electric power generation process, in which a gaseous or liquid fuel is burned in a combustion turbine that both drives electrical generators and provides heat to produce steam in a heat recovery steam generator (HRSG). The steam produced by the HRSG is then fed to a steam turbine that drives a second electric generator. The combination of using the energy released by burning a fuel to drive both a combustion turbine generator set and a stream turbine generator significantly increases the overall efficiency of the electric power generation process.

Coal is the most abundant fossil fuel in the United States and is predominately used for electric power generation. In 2008, approximately 49% of the net electricity generated in the U.S. was produced by coal (U.S. EIA 2010). Historically, electric utilities have burned solid coal in steam generating units. However, coal can also be first gasified and then burned as a gaseous fuel. The integration of coal gasification technologies with the combined cycle electric generation process is called an integrated gasification combined cycle (IGCC) system or a "coal gasification facility". For the remainder of this document, the term "electric generating unit" or "EGU" is used to mean a solid fuel-fired steam generating unit that serves a generator that produces electricity for sale to the electric grid.

2. COAL-FIRED ELECTRIC GENERATING UNITS

This section provides a summary overview of the types or ranks of coal that are typically burned in EGUs operating in the United States, the most commonly used combustion processes, and the resulting emissions of greenhouse gases.

2.1. Coals Burned in U.S. EGUs

In the United States, coals are ranked based on the degree of metamorphism (effectively, the geological age of the coal and the conditions under which the coal formed). These classification criteria have been standardized by the American Society for Testing and Materials (ASTM) method D-388. Under the ASTM method, coals are divided into four major categories called "ranks:" anthracite, bituminous coal, subbituminous coal, and lignite. Typical coal characteristics for the three most commonly used coal ranks are summarized in Exhibit 2-1.

**Exhibit 2-1. Selected characteristics of major coal ranks
used for electricity generation in the United States**

Coal Rank[a]	Higher Heating Value (HHV) Range Defined by ASTM D-388	Typical Coal Moisture Content[b]	Coal Delivered for U.S. Electric Power Production in 2008[c,d]		
			Total Coal Quantity Delivered Nationwide (1,000 tons)	Average Ash Content	Average Sulfur Content
Bituminous	>10,500 Btu/lb	2 to 16%	463,943	10.6%	1.68%
Subbituminous	<10,500 Btu/lb and >8,300 Btu/lb	15 to 30%	522,228	5.8%	0.34%
Lignite	< 8,300 Btu/lb	25 to 40%	68,945	13.8%	0.86%

[a] Anthracite coal use is limited to reclaiming coal from coal refuse piles for use in a few power plants located close to the anthracite mines in eastern Pennsylvania.
[b] Reference: U.S. EPA, 2001.
[c] Reference: U.S. EIA, 2010, Table 3.6.
[d] Includes data collected from electric utilities, independent power producers, and combined heat and power producers.

Most coal-fired EGUs in the United States burn either bituminous or subbituminous coals. Approximately one half of the tonnage of coals delivered to U.S. electric power generation facilities was subbituminous (49.5%), and another 44% was bituminous coal. Some coal-fired EGUs burn multiple coal ranks. At many of these facilities, the coals are blended together before firing. However, some facilities may switch between coal ranks because of site-specific considerations. The largest sources of bituminous coals burned in EGUs are mines in regions along the Appalachian Mountains, in southern Illinois, and in Indiana. Additional bituminous coals are supplied from mines in Utah and Colorado. The vast majority of subbituminous coals are supplied from mines in Wyoming and Montana, and many EGUs burn subbituminous coals from the Powder River Basin (PRB) region in Wyoming. This material is often referred to simply as "PRB coal."

In general, the burning of lignite or anthracite by electric utilities is limited to those EGUs that are located near the mines supplying the coal. Lignite accounted for approximately 6.5% of the total tonnage of coal delivered to electric utility power plants in 2008. All of those facilities were located near the coal deposits from which the lignite was mined in Texas, Louisiana, Mississippi, Montana, or North Dakota. Similarly, anthracite use was limited to a few power plants located close to the anthracite mines in eastern Pennsylvania. The coal-fired EGUs at those facilities primarily burn anthracite that has been reclaimed from coal refuse piles of previous mining operations. In general, "coal refuse" means any by-product of coal mining or coal cleaning operations with an ash content greater than 50 % (by weight) and a heating value less than 13,900 kilojoules per kilogram (kJ/kg) (6,000 Btu per pound (Btu/lb) on a dry basis. Coal refuse piles from previous mining operations are primarily located in Pennsylvania and West Virginia. Current mining operations generate less coal refuse than older ones.

2.2. Coal Utilization in U.S. EGUs

Steam turbine power plants operate on the Rankine thermodynamic cycle. The steam is produced by the boiler, where water pumped into the boiler ("feedwater") passes through a series of tubes to capture heat released by coal combustion and then boils under high pressure to become superheated steam. The superheated steam leaving the boiler then enters the steam turbine throttle, where it powers the turbine and connected generator to make electricity.

After the steam expands through the turbine, it exits the back end of the turbine into the surface condenser, where it is cooled and condensed back to water. This condensate is then returned to the boiler through high-pressure feed pumps for reuse. Heat from the condensing steam is normally rejected to cooling water circulated through the condenser which then goes to a surface water body, such as a river, or to an on-site cooling tower.

An EGU can be classified as either dry or wet bottom, depending on the ash removal technique used. Dry bottom boilers fire coals with high ash fusion temperatures, allowing for solid ash removal. In the less common wet bottom (slag tap) boilers, coal with a low ash fusion temperature is fired, and molten ash is drained from the bottom of the boiler.

Coal-fired EGUs use one of five basic coal utilization processes.

- Stoker-fired
- Pulverized coal (PC)
- Cyclone-fired
- Fluidized-bed combustion (FBC)
- Coal gasification (IGCC)

To improve the overall thermal conversion efficiency of the Rankine cycle, the majority of EGUs include a series of heat recovery sections. These sections are located downstream from the furnace chamber and are used to extract additional heat from the flue gas. The first section contains a "superheater," which is used to increase the steam temperature. The second heat recovery section contains a "reheater," which reheats the steam exhausted from the first stage of the steam turbine. This steam is then returned for another pass thorough a second stage of the turbine. The reheater is followed by an "economizer," which preheats the condensed feedwater recycled back to the boiler tubes in the furnace. The final heat recovery section is the "air heater," which preheats the ambient air used for coal combustion. The flue gas exhausted from the boiler passes through particulate matter (PM) and other air emissions control equipment before being vented to the atmosphere through a stack.

2.2.1. Stoker-Fired Coal Combustion

First introduced to the electric utility industry in the late 1800s, stoker-fired coal combustion is the oldest boiler coal-firing design. In a stoker-fired boiler, the coal is crushed and burned on a grate. Heated air passes upward through openings in the grate. Stokers are classified according to the way coal is fed to the grate – as underfeed stokers, overfeed stokers, and spreader stokers (see Exhibit 2-2). Stoker firing coal combustion is an obsolete technology for new coal-fired EGUs because the other newer coal combustion technologies provide superior coal combustion efficiency, applicability, and other advantages. There are still a few small stoker-fired EGUs in service in the U.S., but as these units are retired no new coal-fired stoker-fired EGUs are expected to be built. The majority of new stoker-fired boiler capacity is expected to occur at municipal solid waste combustor facilities and facilities burning solid biomass.

2.2.2. Pulverized-Coal Combustion

Pulverizing coal into a very fine powder allows the coal to be burned more easily and efficiently. For a PC-fired EGU, the coal must first be pulverized in a mill to the consistency of talcum powder (i.e., at least 70% of the particles will pass through a 200-mesh sieve). The pulverized coal is generally entrained in primary combustion air before being blown through the burners into the combustion chamber where it is fired in suspension. PC-fired boilers are classified by the firing position of the burners either as wall-fired or tangential-fired (see Exhibit 2-2).

A PC-fired boiler consists of multiple sections, and Exhibit 2-3 presents a simplified schematic of the major components of a PC-fired boiler using subcritical steam conditions. The pulverized coal is ignited and burned in the section of the boiler called the "furnace chamber" (or sometimes the "firebox"). Ambient air blown into the furnace chamber provides the oxygen required for combustion. The walls of the furnace chamber are lined with vertical tubes containing the feedwater. Heat transfer from the hot combustion gases in the furnace boils the water in the tubes to produce the high-temperature, high-pressure steam. The steam is separated from boiler water in a steam drum and sent to the steam turbine. The remaining water in the drum re-enters the boiler for further conversion to steam. The hot combustion products are vented from the furnace in a gas stream called collectively flue gas.

Exhibit 2-2. Characteristics of coal-firing configurations used for U.S. EGUs

Coal-firing Configuration	Application to U.S. EGUs	Coal Combustion Process Description	Distinctive Design/Operating Characteristics	
Stoker-fired	• Oldest coal-firing design first introduced to the electric utility industry in the late 1800s. • Not a significant contributor to overall U.S. nationwide MW generating capacity. • New EGUs are not expected to use this coal-firing design because of the superior performance and advantages of newer coal combustion technologies.	Coal is crushed into large lumps and burned in a fuel bed on a moving, vibrating, or stationary grate. Coal is pushed, dropped, or thrown onto the grate by a mechanical device called a "stoker."	Spreader-stoker	A flipping mechanism throws the coal into the furnace above the grate. The fine coal particles burn in suspension while heavier coal lumps fall to the grate and burn in a fuel bed.
			Underfeed	Coal fed by pushing the coal up underneath the burning fuel bed.
			Traveling grate	Coal is fed by gravity onto a moving grate and leveled by a stationary bar at the furnace entrance.
Pulverized-Coal Combustion	• Coal-firing design predominately used at existing U.S. EGUs • In 2008, consumed ~ 92% of total coal consumed by U.S. EGUs.[a] • Currently coal-firing design of choice for new large coal-fired EGUs (> 400 MWe) built in U.S.	Coal is ground to a fine powder that is pneumatically fed to a burner where it is mixed with combustion air and then blown into the furnace. The pulverized-coal particles burn in suspension in the furnace. Unburned and partially burned coal particles are carried off with the flue	Wall-fired	An array of burners fire into the furnace horizontally, and can be positioned on one wall or opposing walls depending on the furnace design.
			Tangential-fired (Corner-fired)	Multiple burners are positioned in opposite corners of the furnace producing a fireball that moves in a cyclonic motion and expands to fill the furnace.
Cyclone	• Existing cyclone EGUs in U.S. constructed prior to 1981. • In 2008, consumed ~ 6% of total coal consumed by U.S. EGUs. • New EGUs are not expected to use this boiler type because of the commercial availability of FBC technology.	Coal is crushed into small pieces and fed through a burner into the cyclone furnace. A portion of the combustion air enters the burner tangentially creating a whirling motion to the incoming coal.	Designed to burn coals with low-ash fusion temperatures that are difficult to burn in PC boilers. The majority of the ash is retained in the form of a molten slag.	

Exhibit 2-2. (Continued)

Coal-firing Configuration	Application to U.S. EGUs	Coal Combustion Process Description	Distinctive Design/Operating Characteristics	
Fluidized-bed Combustion	• FBC EGUs increasingly being built in U.S. to burn low rank coals, coal refuse, and blends of coal with other solid fuels such as petroleum coke or biomass. • In 2008, consumed approximately 2% of total coal consumed by U.S. EGUs.a • Atmospheric FBC EGUs are currently operating in the U.S. with generating capacities in the range of 250 to 300 MWe. • No Pressurized FBC boilers currently used for U.S. EGUs	Coal is crushed into fine particles. The coal particles are suspended in a fluidized bed by upward-blowing jets of air. The result is a turbulent mixing of combustion air with the coal particles. Typically, the coal is mixed with a sorbent such as limestone (for SO_2 emission control). The unit can be designed for combustion within the bed to occur at atmospheric or elevated pressures. Operating temperatures for FBC are in the range of 1,500 to 1,650°F (800 to 900°C).	Bubbling fluidized bed (BFB)	Operates at relatively low gas stream velocities and with coarse-bed size particles. Air in excess of that required to fluidize the bed passes through the bed in the form of bubbles.
			Circulating fluidized bed (CFB)	Operates at higher gas stream velocities and with finer-bed size particles. No defined bed surface. Must use high-volume, hot cyclone separators to recirculate entrained solid particles in flue gas to maintain the bed and achieve high combustion efficiency.
Coal Gasification (e.g., IGCC)	• Limited application to EGUs to date. • Some new proposed EGU projects using coal gasification as part of IGCC plant.	Synthetic combustible gas ("syngas") derived from an on-site coal gasification process is burned in a combustion turbine. The hot exhaust gases from the combustion turbine pass through a heat recovery steam generator to produce steam for driving a steam turbine/generator unit.	Coal gasification units are unique from the other coal-firing configurations because a gaseous fuel (synfuel or syngas) is burned instead of solid coal and combines the Rankine and Brayton thermodynamic cycles as is the case for a combined cycle power plant.	

a Source: U.S. EIA, 2008.

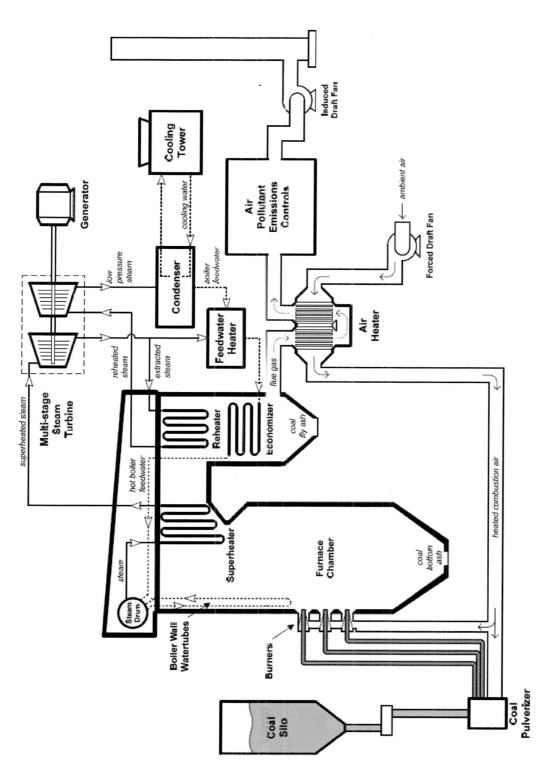

Exhibit 2-3. Simplified schematic of a PC-fired EGU using a subcritical boiler.

2.2.3. Cyclone Coal Combustion

Cyclone coal combustion technology was developed as an alternative to PC-firing because it requires less pre-processing of the coal and allows for the burning of lower rank coals with higher moisture and ash contents. Cyclone boilers use burner design and placement (i.e., several water-cooled horizontal burners) to produce high-temperature flames that circulate in a cyclonic pattern. The coal is crushed to a 4-mesh size, and then fed tangentially with primary air, to a horizontal cylindrical combustion chamber. In this chamber, small coal particles are burned in suspension, while the larger particles are forced against the outer wall. The high temperatures developed in the relatively small boiler volume, combined with the low fusion temperature of the coal ash, causes the ash to form a molten slag, which is drained from the bottom of the boiler through a slag tap opening. Existing cyclone EGUs in the U.S. were designed or installed before 1981. Cyclone EGUs have high nitrogen oxides (NOx) emission rates and no new cyclone boilers are expected to be built. Fluidized-bed combustion is an alternative technology that is able to burn lower rank coals without high NOx emissions.

2.2.4. Fluidized-Bed Combustion

The term "fluidized" refers to the state of the bed materials (fuel and inert material [or sorbent]) as gas passes through the bed. In a typical FBC EGU, combustion occurs when coal and a sorbent, such as limestone, are suspended through the action of primary combustion air distributed below the combustor floor. The gas cushion between the solids allows the particles to move freely, giving the bed a liquid-like characteristic (i.e., fluidized). FBC can occur in either atmospheric or pressurized boilers. Two fluidized bed designs can be used for atmospheric and pressurized FBC boilers: a bubbling fluidized bed or a circulating fluidized bed (CFB) (see Exhibit 2-2). An advantage of CFB boiler EGUs compared to PC-fired EGUs is fuel flexibility. A CFB boiler EGU can burn any rank of coal (including coal refuse), petroleum coke (a carbonaceous solid derived from oil refinery coker units or other cracking processes), and biomass without significant modifications.

The combustion temperature of a FBC boiler (1,500 to 1,650°F) is significantly lower than a PC-fired boiler (2,450 to 2,750°F), which results in lower NOX formation and the ability to capture sulfur dioxide (SO_2) with limestone injection in the furnace. Even though the combustion temperature of a FBC boiler is low, the circulation of hot particles provides efficient heat transfer to the furnace walls and allows longer residence time for carbon combustion and limestone reaction. This results in good combustion efficiencies, comparable to PC-fired EGUs.

Atmospheric CFB boilers have successfully been scaled-up and are operating at a number of facilities throughout the world. Exhibit 2-4 presents a simplified schematic of the major components of a CFB boiler EGU. Calcium in the sorbent combines with SO_2 gas to form calcium sulfite and sulfate solids, and solids exit the combustion chamber and flow into a hot cyclone. The cyclone separates the solids from the gases, and the solids are recycled for combustor temperature control. Heat in the flue gas exiting the hot cyclone is recovered in a series of heat recovery sections of the boiler to produce steam. The superheated steam leaving the boiler then enters the steam turbine, which powers a generator to produce electricity. Like PC-fired EGUs, CFB boilers can be used with either subcritical or supercritical steam cycles.

Currently, the capacity of CFB subcritical boilers ranges from 25 to 350 MWe. Examples of these systems include (Foster Wheeler North America Corp., 2009):

- Two 300 MWe CFB subcritical boilers at the Jacksonville Energy Authority power plant in Jacksonville, Florida. These units are capable of burning either 100% coal or 100% petroleum coke or any combination of the two.
- Three 262 MWe CFB subcritical boilers at the Turow power plant in Poland. The fuel for these boilers is lignite with moisture content of 45% by weight.

The largest atmospheric CFB boiler in operation to date is a 460 MWe unit at a power plant owned by the Polish utility company Południowy Koncern Energetyczny SA (PKE) in Lagisza, Poland (Foster Wheeler North America Corp., 2009). This unit is also the world's first supercritical CFB boiler. The primary fuel burned in the unit is Polish bituminous coal. The commercial operation of this unit demonstrates the successful integration of CFB boiler technology with supercritical boiler technology. The unit features include a vertical evaporator with supercritical steam conditions (4,000 psia, 1,050/1,075°F) and a reported overall net plant efficiency of 41.6% (HHV basis). Based on the design and operating experience with the Lagisza Power Plant, both 600 and 800 MWe size supercritical CFB boiler designs with full commercial guarantees are being offered (Foster Wheeler North America Corp., 2009).

Pressurized fluidized-bed combustion (PFBC) systems are FBC systems that operate at elevated pressures (typically pressures of 1-1.5 MPa) and produce a high-pressure gas stream at temperatures that can drive a turbine. As with atmospheric FBC, two formats are possible, one with bubbling beds, the other with a circulating configuration. Currently, all operating units use bubbling beds. In a PFBC, the combustor and hot gas cyclones are all enclosed in a pressure vessel. Both coal and sorbent (for SO_2 emissions reductions) have to be fed across the pressure boundary, and similar provision for ash removal is necessary. For hard coal (i.e., bituminous coal) applications, the coal and limestone can be crushed together, and then fed as a paste, with 25% water. As with atmospheric FBC, a combustion temperature between 1,500 to 1,650°F (800 to 900°C) has the advantage of less NOX formation than in PC combustion. In addition, the effectiveness of a CCS system is increased due to the high pressure within the PFBC cycle and higher partial pressure of the CO_2 in the hot gas stream.

The initial or first-generation PFBC designs are based on directly burning crushed coal in the combustor. The high pressure gas is first expanded through a turbine and then heat is recovered from the turbine exhaust in a HRSG to produce steam, which is used to drive a conventional steam turbine. Exhibit 2-5 presents a simplified schematic of the major components of a PFBC EGU. A number of demonstration projects (ranging in size from 60 to 130 MWe) were conducted during the 1990s in Japan, Spain, Sweden, the United Kingdom, the U.S., and other countries. Japanese equipment manufacturers and electric power companies have led the commercial development of PFBC technology with the construction of several commercial-scale units.

- 360 MWe PFBC unit operated by Kyushu Electric Power Company at the Karita Power Station located near Kitakyushu, Japan. The unit began commercial operation in July 2001. The unit uses a supercritical boiler and has a reported net efficiency based on test results of 41.8% HHV (Asai, 2004).
- 250 MWe PFBC unit operated by Chugoku Electric Power Co., Inc. at Osaki Power Station located near Hiroshima, Japan. Unit 1 began commercial operation in 2000. The planned construction of a second PFBC unit at the facility was cancelled in 2008

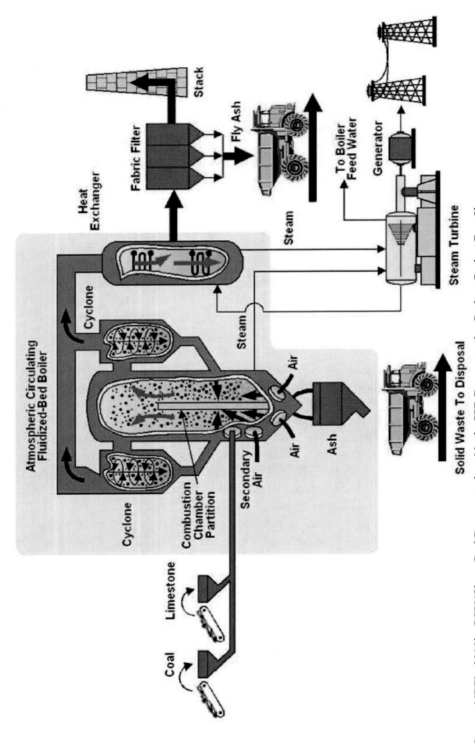

Source: NETL, 2010b, CCPI/Clean Coal Demonstrations Nucla CFB Demonstration Project, Project Fact Sheet.

Exhibit 2-4. Simplified schematic of an atmospheric circulating fluidized-bed (CFB) boiler power plant.

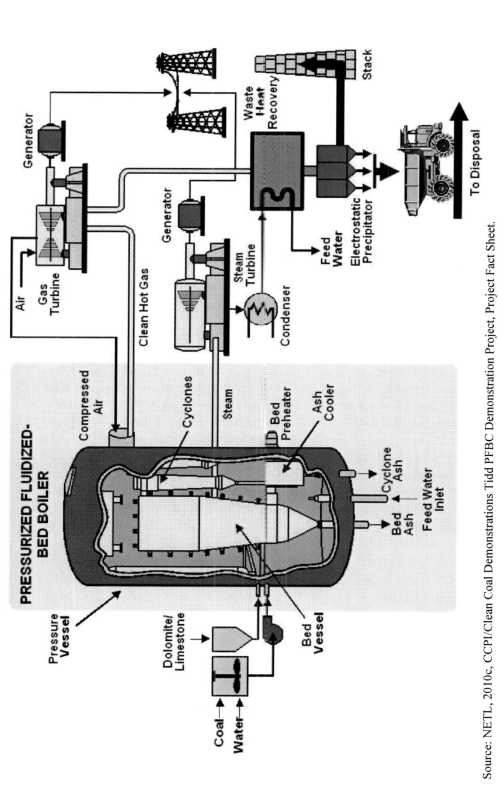

Source: NETL, 2010c, CCPI/Clean Coal Demonstrations Tidd PFBC Demonstration Project, Project Fact Sheet.

Exhibit 2-5. Simplified schematic of a pressurized fluidized-bed combustion (PFBC) power plant.

More advanced second-generation PFBC system designs use a pressurized carbonizer to first process the feed coal into fuel gas and char (solid material that remains after light gases and tar have been driven-out during the initial stage of combustion). The PFBC burns the char to produce steam and to heat combustion air for the combustion turbine. The fuel gas from the carbonizer burns in a topping combustor linked to a combustion turbine, heating the gases to the rated firing temperature of the combustion turbine. Heat is recovered from the combustion turbine exhaust in a HRSG to produce steam, which is used to drive a conventional steam turbine. These systems are also called advanced circulating pressurized fluidized-bed combustion (APFBC) combined cycle systems.

2.2.5. Coal Gasification

An IGCC power plant uses a coal gasification system to convert coal into a synthetic gas, which is then used as fuel in a combined cycle electric generation process. Coal is gasified by a process in which coal or a coal/water slurry is reacted at high temperature and pressure with oxygen (or air) and steam in a vessel referred to as a "gasifier" to produce a combustible gas composed of a mixture of carbon monoxide (CO) and hydrogen. This gas is often referred to as synthetic gas or syngas. Gasification processes have been developed using a variety of designs including moving bed, fluidized bed, entrained flow, and transport gasifiers. Coal gasification processes are offered by a number of companies with varying degrees of existing commercial application (NETL, 2010a). Exhibit 2-6 presents a simplified schematic of the major components of an IGCC power plant. The hot syngas can then be processed to remove sulfur compounds, mercury, and PM before it is used to fuel a combustion turbine generator to produce electricity. The heat in the exhaust gases from the combustion turbine is recovered to generate additional steam. This steam, along with the steam produced by the gasification process, then drives a steam turbine generator to produce additional electricity.

The efficiency of an IGCC power plant is comparable to the latest advanced PC-fired and CFB EGU designs using supercritical boilers. The advantages of using IGCC technology can include greater fuel flexibility (e.g., capability to use a wider variety of coal ranks), potential improved control of PM, SO_2 emissions, and other air pollutants, with the need for fewer post-combustion control devices (e.g., almost all of the sulfur and ash in the coal can be removed once the fuel is gasified and prior to combustion), generation of less solid waste requiring disposal, and reduced water consumption when compared to an EGU using a supercritical boiler (U.S. EPA, 2006). Disadvantages of using IGCC include additional plant complexity, higher construction costs, and poorer performance at high altitude locations when compared to an EGU using a supercritical boiler. However, IGCC power plants offer the potential for lower control costs of CO_2 emissions because the CO_2 in the syngas can be removed prior to combustion. Interest by U.S. electric utilities in building new IGCC power plants is increasing because of site-specific considerations and potential cost benefits for the technology.

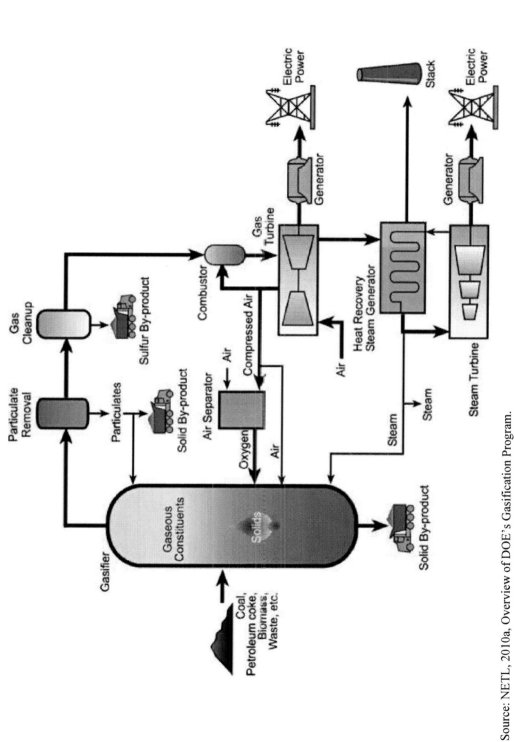

Source: NETL, 2010a, Overview of DOE's Gasification Program.

Exhibit 2-6. Simplified schematic of an integrated gasification combined cycle (IGCC) power plant.

Currently operating IGCC plants include the following U.S. and foreign plants (NETL, 2010a):

- 253 MWe IGCC plant at the NUON Willem-Alexander Power Plant in Buggenum, Netherlands. The unit began operation in 1994.
- 262 MWe IGCC plant at the Duke Energy Wabash River Power Station in Indiana. The unit began operation in 1995.
- 250 MWe IGCC plant at the Tampa Electric Company (TECO) Polk Power Station in Florida. The unit began operation in 1996.
- 400 MWe IGCC plant at the SUV power plant in Vresova, Czech Republic. The unit began operation in 1996.
- 283 MWe IGCC plant at the ELCOGAS power plant in Puertollano, Spain. The unit began operation in 1998.
- 250 MWe IGCC plant at the Joban Joint Electric Power Company Nakoso Power Station in Iwaki City, Japan. The unit began operation in 2007.

Over the past 5 years, a number of larger IGCC power plant projects have been proposed by U.S. electric utility companies. Some of these IGCC projects have been indefinitely delayed or canceled because of economic and regulatory factors, such as escalating project investment costs beyond initial estimates and unresolved cost recovery issues with State public utility commissions. One commercial IGCC project currently under construction is a 630 MWe IGCC facility at the Duke Energy Edwardsport Power Station in Knox County, Indiana.

Syngas produced by coal gasification can not only be used as a fuel to generate electricity or steam but also as a basic chemical building block for a large number of petrochemical and refining products. Because of these multiple uses, future IGCC projects may include facilities that integrate electricity generation with the production of other industrial outputs such as chemical feedstocks for manufacturing operations or hydrogen fuel for vehicles.

2.3. GHG Emissions from Coal-Fired EGUs

The principal chemical constituents of coal are carbon, hydrogen, oxygen, nitrogen, sulfur, moisture, and incombustible mineral matter (i.e., ash). When coal is burned, the carbon and hydrogen are oxidized to form the primary combustion products of CO_2 and water. Other combustion products such as NOX, SO_2, CO, and PM are formed in varying amounts.

The principal GHGs that enter the atmosphere because of human activities are CO_2, nitrous oxide (N_2O), methane (CH4), hydrofluorocarbons (HFC's), perfluorocarbons (PCF's), and sulfur hexafluoride (SF6). Of these, CO_2 is by far the most abundant GHG emitted from power production by coal utilization. To optimize overall efficiency for a given EGU, the unit is operated under conditions such that nearly all of the fuel carbon is converted to CO_2 during the combustion process. Methane is emitted during the mining and transport of coal but is not a significant by-product of EGU coal combustion. Fluorinated gases are not formed by coal combustion. Sulfur hexafluoride might be used at the power plant switchyard, but the switchyard is not typically considered part of the EGU.

Formation of N_2O during the combustion process results from a complex series of reactions and its formation is dependent upon many factors. However, the formation of N_2O is minimized when combustion temperatures are kept high and excess air is kept to a minimum. PC-fired EGUs are typically operated at these conditions and are not significant sources of N2O emissions. However, FBC EGUs can have measurable N2O emissions, resulting from the lower combustion temperatures and the use of selective noncatalytic reduction (SNCR) to reduce NOX emissions. Operating factors impacting N2O formation include combustion temperature, excess air, and sorbent feed rate (Korhonen, 2001). The N2O formation resulting from SNCR depends upon the reagent used, the amount of reagent injected, and the injection temperature (Weijuan, 2007).

2.4. Factors Impacting Coal-Fired EGU CO_2 Emissions

The level of CO_2 emissions that can potentially be released from a given coal-fired EGU depends on the type of coal burned, the overall efficiency of the power generation process, and use of air pollution control devices

2.4.1. Impact of Coal Rank on CO2 Emissions from EGUs

The amount of CO_2 that potentially can be emitted from a coal-fired EGU varies depending on the coal rank burned. The amount of heat released by coal combustion depends on the amounts of carbon, hydrogen, and oxygen present in the coal and, to a lesser extent, on the sulfur content. Hence, the ratio of carbon to heat content depends on these heat-producing components of coal, and these components vary by coal rank. **Exhibit 2-7** presents a comparison of the CO_2 emissions for the average heating values of U.S. coals. The values presented in the table are arithmetic averages and assume complete combustion. Based on these averages, in general anthracite emits the largest amount of CO_2 per million Btu (MMBtu), followed by lignite, subbituminous coal, and bituminous coal. However, for a given coal rank there is variation in the CO_2 emission factor depending on the coal bed from which the coal is mined.

Exhibit 2-7. CO_2 emission factors for coal by coal rank

Coal Rank	CO_2 Emissions per Unit of Heat Input (lbs CO_2/MMBtu)	
	U.S. Average	Range Across States with Coal Rank Deposits
Anthracite	227.4	227.4
Bituminous	205.3	201.3 to 211.6
Subbituminous	211.9	207.1 to 214.0
Lignite	216.3	211.7 to 220.6

Source: U.S. EIA (Hong, R. and E. Slatick, 1994).

In addition to the lower CO_2 emissions rate per unit of heat input (lbs CO_2/MMBtu), due to the inherent moisture in subbituminous and lignite coals, all else being equal a bituminous coal-fired boiler is more efficient than a corresponding boiler burning subbituminous or lignite coal. Therefore, switching from a low to a high-rank coal will tend to lower GHG emissions from the utility stack. However, overall GHG emissions might not be lowered by switching to bituminous coal. All coal mining operations release coal bed methane to the atmosphere during the mining process. Some bituminous coal reserves release significant amounts of methane, which could, in theory, offset GHG savings. Additional factors when considering overall GHG emissions include the fuel needs to mine, process, and transport the coal.

Additional solid fuels burned in EGUs include petroleum coke, biomass, and municipal solid waste (MSW). Petroleum coke has one of the highest CO_2 emissions rate (225 lb CO_2/MMBtu) of commonly used solid fuels. MSW combustors provide significant GHG reductions as an alternate to landfills (Kaplan, 2008). However, due to the difficulties associated with transporting large amounts of solid waste, MSW combustor facilities used for electrical power generation are typically limited to less than 100 MW of electrical output.

Of the gaseous and liquid fossil fuels used in steam generating units, natural gas combustion releases approximately 117 pounds of CO_2 per MMBtu, distillate oil releases 161 lb CO_2/MMBtu, and residual oil releases 174 lb CO_2/MMBtu. However, none of these fuels are typically used in new baseload steam generating units (e.g., boilers). Natural gas and distillate oil are significantly more expensive per unit heat input than coal. In addition, combustion turbines burning natural gas and distillate oil generate power more efficiently than a boiler burning natural gas and distillate oil. New baseload electric generation based on the use of either natural gas or distillate would likely use combined cycle combustion turbines. Therefore, aside from small amounts of natural gas for startup, shutdown, and potentially for combustion control, few new steam generating units are expected to burn significant quantities of either of these fuels directly in the boiler. Existing EGUs that burn natural gas and distillate oil tend to be older units that operate in a peaking or cycling mode. However, several base load coal-fired EGUs have been converted to natural gas. Natural gas-fired boilers tend to be less efficient than coal-fired boilers; however, they can startup and change loads more quickly than similar coal-fired boilers, do not typically require post combustion controls, and the fuel handling is simpler. Residual oil also tends to be more expensive than coal per unit of heat input, and because post-combustion environmental controls would still often be required, it is also not a common fuel choice for EGUs in the Lower 48 States. There has not been a new residual oil-fired EGU built in the Lower 48 States since 1981.

2.4.2 Impact of Coal-Fired EGU Efficiency on CO_2 Emissions

As the thermal efficiency of a coal-fired EGU is increased, less coal is burned per kilowatt-hour (kWh) generated, and there is a corresponding decrease in CO_2 and other air emissions. There is no standardized procedure for continuous on-line measurement of coal-fired EGU thermal efficiency (Peltier, 2010). However, a near approximation performed under EPA's Acid Rain Program collects heat input and gross megawatt output on an hourly basis to calculate gross heat rate. The heat input is derived from standardized continuous emission monitors, while the utility supplies gross megawatt output. The electric energy output as a fraction of the fuel energy input expressed in percentage is a commonly-used practice for reporting the efficiency of a coal-fired EGU. The greater the output of electric energy for a given amount of fuel energy input, the higher the efficiency for the electric generation process. Heat rate is another common way to express efficiency. Heat rate is expressed as the number of Btu or kJ required to generate a kWh of electricity. Lower heat rates are associated with more efficient power generating plants. Although the same basic formula is used to calculate efficiency for coal-fired EGUs, there are different methodologies for measuring the appropriate parameters. For example, the varying accuracy of the different methodologies can cause discrepancies in the measurement the heating value of the coal burned.

Efficiency can be calculated using the higher heating value (HHV) or the lower heating value (LHV) determined for the fuel. The HHV is the heating value directly determined by calorimetric measurement of the fuel in the laboratory. The LHV is calculated using a formula to account for the moisture in the fuel (i.e., subtract the energy required to vaporize the water in the coal and is thus not available to produce steam) and is a smaller value than the HHV. Consequently, the HHV efficiency for a given EGU is always lower than the corresponding LHV efficiency, because the reported heat input is larger. For bituminous coals the HHV efficiency value is typically about 2 percentage points lower than the corresponding LHV efficiency. For higher moisture subbituminous coals and lignites, the HHV efficiency is approximately 3 to 5 percentage points lower than the corresponding LHV efficiency (depending on moisture content). In engineering practice, HHV is typically used in the U.S. to express the efficiency of steam electric power plants while in Europe the practice is to use LHV.

Similarly, the electric energy output for an EGU can be expressed as either of two measured values. One value relates to the amount of total electric power generated by the EGU, or "gross output." However, a portion of this electricity must be used by the EGU facility to operate the unit, including pumps, fans, electric motors, and pollution control equipment. This in-facility electrical load, often referred to as the "parasitic load," reduces the amount of power that can be delivered to the transmission grid for distribution and sale to customers. Consequently, electric energy output is also expressed in terms of "net output," which reflects the EGU gross output minus its parasitic load.

When using efficiency to compare the effectiveness of different coal-fired EGU configurations and the applicable GHG emissions control technologies, it is important to ensure that all efficiencies are calculated using the same type of heating value (i.e., HHV or LHV) and the same type of electric energy output (i.e., gross MWh or net MWh).

Although there is a direct inverse correlation between coal-fired EGU efficiency and CO_2 emissions, other factors must be considered when comparing the effectiveness of GHG control technologies to improve the efficiency of a given coal-fired EGU. The actual overall efficiency that a given coal-fired EGU achieves is determined by the interaction of a combination of site-specific factors that impact efficiency to varying degrees. These factors include:

- *EGU thermodynamic cycle* – EGU efficiency can be significant improved by using a supercritical or ultra-supercritical steam cycle.
- *EGU coal rank and quality* – EGUs burning higher quality coals (e.g., bituminous) tend to be more efficient than EGUs burning lower quality coals (e.g., lignite).
- *EGU plant size* – The electric-generating capacity of EGUs ranges from approximately 25 to 1,300 MWe. Assuming an EGU efficiency of 33% (a typical efficiency for existing coal-fired EGUs), this corresponds to a heat input range of 250

to 13,400 MMBtu/hr. EGU efficiency generally increases with size because the boiler and steam turbine losses are lower for larger equipment. However, as equipment size increases the differences in these losses start to taper off.

- *EGU pollution control systems* – The electric power consumed by air pollution control equipment reduces the overall efficiency of the EGU.
- *EGU operating and maintenance practices* – The specific practices used by an individual electric utility company for combustion optimization, equipment maintenance, etc. can affect EGU efficiency.
- *EGU cooling system* – The temperature of the cooling water entering the condenser can have impacts on steam turbine performance. Once-through cooling systems can have an efficiency advantage over recirculating cooling systems (e.g., cooling towers). However, once-though cooling systems typically have larger water related ecological concerns than recirculating cooling systems.
- *EGU geographic location* – The elevation and seasonal ambient temperatures at the facility site potentially may have a measureable impact on EGU efficiency. At higher elevations, air pressure is lower and less oxygen is available for combustion per unit volume of ambient air than at lower elevations. Cooler ambient temperatures theoretically could increase the overall EGU efficiency by increasing the draft pressure of the boiler flue gases and the condenser vacuum, and by increasing the efficiency of a condenser recirculating cooling system.
- *EGU load generation flexibility requirements* – Operating an EGU as a baseload unit is more efficient than operating an EGU as a load cycling unit to respond to fluctuations in customer electricity demand.
- *EGU equipment manufacturers* – The efficiency specifications of major EGU components such as boilers, turbines, and electrical generators provided by equipment manufacturers can affect EGU efficiency.
- *EGU plant components* – EGUs using the optimum number of feedwater heaters, high-efficiency electric motors, variable speed drives, better materials for heat exchangers, etc. tend to be more efficient.

Because of these factors, coal-fired EGUs that are identical in design but operated by different utility companies in different locations may have different efficiencies. Thus, the level of effectiveness of a given GHG control technology used to improve the efficiency at one coal-fired EGU facility may not necessarily directly transfer to a coal-fired EGU facility at a different location.

2.4.3. Impact of SO_2 Controls on Coal-Fired EGU CO_2 Emissions

The SO_2 emissions from new coal-fired EGUs, or retrofitting of an existing facility without specific SO_2 controls, are controlled using flue gas desulfurization (FGD) technology to remove the SO_2 before it is vented to the atmosphere. The selection of the type of FGD technology will impact overall GHG emissions. All FGD processes require varying amounts of electric power to operate, which contributes to the overall parasitic load of the unit. The FGD parasitic load requirements are typically between 1-2% of the gross output of the facility. In addition, some FGD processes use carbon-containing reagents (e.g., carbonates) that form CO_2 as a byproduct of the chemical reactions of the reagent with SO_2.

Exhibit 2-8. CO$_2$ formation from coal-fired EGU flue gas desulfurization (FGD) processes

FGD Type	Reagent	Forms CO$_2$	Overall Reaction(s)	Reference
Wet Scrubbing	Limestone (CaCO$_3$)	yes	$CaCO_3 + SO_2 + \frac{1}{2}H_2O \rightarrow CaSO_3 \cdot \frac{1}{2}H_2O(s) + CO_2$; $CaCO_3 + SO_2 + 2H_2O + \frac{1}{2}O_2 \rightarrow CaSO_4 \cdot 2H_2O(s) + CO_2$	Ref 1
	Magnesium-enhanced lime; Dolomitic lime [Ca(OH)$_2$•Mg(OH)$_2$]	no	$10Ca(OH)_2 + 11SO_2 + Mg(OH)_2 \rightarrow 10CaSO_3 \cdot \frac{1}{2}H_2O(s) + MgSO_3 + 6H_2O$	Ref 1,4
	Dual Alkali; Sodium solution and lime	no	$2NaOH + SO_2 \rightarrow Na_2SO_3 + H_2O$; $H_2O + Na_2SO_3 + SO_2 \rightarrow 2NaHSO_3$; $2NaHSO_3 + Ca(OH)_2 \rightarrow Na_2SO_3 + CaSO_3 \cdot \frac{1}{2}H_2O + 3/2H_2O$; $Na_2SO_3 + Ca(OH)_2 \rightarrow 2NaOH + CaSO_3$	Ref 2
	Dual Alkali (Dowa)	yes	$Al_2O_3 \cdot Al_2(SO4)_3 + 3SO_2 + 3/2O_2 \rightarrow 2Al_2(SO4)_3$; $2Al_2(SO4)_3 + 3CaCO_3 \rightarrow Al_2O_3 \cdot Al_2(SO4)_3 + 2CaSO_4(s) + 3CO_2$	Ref 3
	Seawater	yes	$2NaHCO_3 + SO_2 \rightarrow Na_2SO_3 + 2CO_2 + H_2O$; $Na_2SO_4 + \frac{1}{2}O_2 \rightarrow Na_2SO_4$	Ref 3
	Magnesium oxide (MgO)	no	$MgO + SO_2 \rightarrow MgSO_3$	Ref 7
	Hydrogen Peroxide (H$_2$O$_2$)	no	$H_2O_2 + SO_2 \rightarrow H_2SO_4$	Ref 3
	Sodium hydroxide (NaOH)	no	$2NaOH + SO_2 \rightarrow Na_2SO_3 + H_2O$; $Na_2SO_3 + \frac{1}{2}O_2 \rightarrow Na_2SO_4$	Ref 3
Dry/Semi-dry Scrubbing	Hydrated calcitic lime (Ca(OH)$_2$)	no	$Ca(OH)_2 + SO_2 \rightarrow CaSO_3 \cdot \frac{1}{2}H_2O(s) + \frac{1}{2}H_2O$; $Ca(OH)_2 + SO_2 + H_2O + \frac{1}{2}O_2 \rightarrow CaSO_4 \cdot 2H_2O(s)$	Ref 1
	Sodium bicarbonate (NaHCO$_3$)	yes	$2NaHCO_3 + SO_2 \rightarrow Na_2SO_3 + 2CO_2 + H_2O$; $2NaHCO_3 + SO_2 + \frac{1}{2}O_2 \rightarrow Na_2SO_4 + 2CO_2 + H_2O$	Ref 8
	Sodium sesquicarbonate (trona)	yes	$2(Na_2CO_3 \cdot NaHCO_3 \cdot 2H_2O) + 3SO_2 \rightarrow 3Na_2SO_3 + 5H_2O + 4CO_2$; $2(Na_2CO_3 \cdot NaHCO_3 \cdot 2H_2O) + 3SO_2 + 3/2O_2 \rightarrow 3Na_2SO_4 + 5H_2O + 4CO_2$	Ref 8
	Sodium carbonate (soda ash, Na$_2$CO$_3$)	yes	$Na_2CO_3 + SO_2 + \frac{1}{2}O_2 \rightarrow Na_2SO_4 + CO_2$	Ref 8
	Pulverized limestone	yes	$CaCO_3 + SO_2 + 2H_2O + \frac{1}{2}O_2 \rightarrow CaSO_4 \cdot 2H_2O(s) + CO_2$	Ref 1
Other Processes	Ammonia (NH$_3$)	no	$SO_2 + 2NH_3 + H_2O + \frac{1}{2}O_2 \rightarrow (NH_4)2SO_4$	Ref 1,5
	Activated carbon	no	$SO_2 + H_2O + \frac{1}{2}O_2 \rightarrow H_2SO_4$	Ref 6

Ref 1. Srivastava, R., W. Jozewicz, and C. Singer, 2001.
Ref 2. Srivastava, R. and W. Jozewicz , 2001.
Ref 3. Davenport, 2006. Ref 4. Benson, 2003. Ref 5. He, 2002.
Ref 6. EPA, 2005. Ref 7. Shand, 2009. Ref 8. Maziuk, 2002.

For a typical unit, the CO_2 that is chemically created in a scrubber adds an additional 1% to the overall GHG emissions, but it can be as high as 3% for facilities burning high sulfur coals. However, from an overall GHG emissions standpoint the use of FGD technologies that do not form byproduct CO_2, such as lime-based scrubbers, do not necessarily reduce emissions. Lime is manufactured by heating limestone in the absence of oxygen to remove a molecule of CO_2 ($CaCO_3$ + heat $\rightarrow$ CaO + CO_2). Unless the CO_2 is sequestered at the lime production facility, overall GHG emissions will be similar. A list of FGD processes used for controlling SO_2 emissions from coal-fired EGUs is presented in Exhibit 2-8, identifying those processes that chemically form additional CO_2.

3. COAL-FIRED EGU CO_2 CONTROL TECHNOLOGIES

The development of effective and commercially viable CO_2 control technologies for coal-fired EGUs is receiving widespread attention from the utilities, technology providers, and government agencies. Some CO_2 control technologies are still in the research and development phase and are not yet ready for commercial application. Other CO_2 control technologies are being demonstrated at larger scales and are progressing towards commercial viability. This remains an active area of research and new projects and technology advances are reported routinely. The discussions of CO_2 mitigation technologies and options presented in this section are based on the development status of a given technology as described in publicly available information as of May 2010.

3.1. Coal-Fired EGU CO_2 Emissions Control Approaches

A number of technologies lowering CO_2 emissions from coal-fired EGUs are currently commercially available or under development. These control measures use one of two basic approaches to reduce the amount of CO_2 released to the atmosphere: 1) by reducing the amount of fuel used (and the amount of CO_2 formed) by improving the energy efficiency of the electrical generation process, or 2) by separating the CO_2 for long-term storage using carbon capture technology.

3.1.1. Efficiency Improvements

When the efficiency of the power generation process is increased, less coal is burned to produce the same amount of electricity. This provides the benefits of lower fuel costs and reduced air pollutant emissions (including CO_2). A number of energy efficiency technologies are available for application to both existing and new coal-fired EGU projects that can provide incremental step improvements to the overall thermal efficiency. The energy efficiency technologies with the potential to achieve the greatest improvements in electric power generation efficiency involve EGU design, equipment selection, and cost decisions that are typically incorporated during the planning and engineering design phases for a new EGU project.

3.1.2. Carbon Capture and Storage

Carbon capture and storage (CCS) involves the separation and capture of CO_2 from flue gas, or syngas in the case of IGCC. It also requires pressurization of the captured CO_2, transportation via pipeline if necessary, and injection and long-term geologic storage. Several different technologies, at varying stages of development, may be considered for the CO_2 separation and capture. Some have been demonstrated at the slip-stream or pilot-scale, while many others are still at the bench-top or laboratory stage of development.

Development of commercially viable processes for capturing CO_2 from EGUs is being funded by U.S. DOE, electric utility companies, and other organizations. These processes typically use solvents, solid sorbents, and membrane-based technologies for separating and capturing CO_2. Amine-based solvent systems are in commercial use for scrubbing CO_2 from industrial flue gases and process gases. However, solvents have yet to be applied to removing the large volumes of CO_2 that would be required for a coal-fired EGU. Solid sorbents can be used to capture CO_2 through chemical adsorption, physical adsorption, or a combination of the two effects. Membrane-based capture uses permeable or semi-permeable materials that allow for the selective transport/separation of CO_2. Oxy-combustion uses high-purity oxygen (O_2) instead of air to combust coal, producing a highly concentrated CO_2 stream that does not require a separation/capture step.

Once the CO_2 is captured, it is transported, if necessary, and stored. Geologic formations such as oil and gas reservoirs, unmineable coal seams, and underground saline formations are potential options for long-term storage. Basalt formations and organic rich shales are also being investigated for potential use as storage. Beneficial reuse (e.g., enhanced oil recovery or carbonation) is a potential alternative to strict storage that provides potential revenue to offset a portion of the CCS costs.

One recent study prepared for the U.S. DOE by the Pacific Northwest National Laboratory (PNNL, 2009) evaluated the development status of various CCS technologies. The study addressed the availability of capture processes; transportation options (CO_2 pipelines); injection technologies; and measurement, verification, and monitoring technologies. The study concluded that, in general, CCS is technically viable today. However, full-scale carbon separation and capture systems have not yet been installed and fully integrated at an EGU. The study also did not address the cost or energy requirements of implementing CCS technology. For up-to-date information on Department of Energy's National Energy Technology Laboratory's (NETL) Carbon Sequestration Program go to the NETL web site at: http://www.netl.doe.gov/technologies/carbon_seq/.

In 2010, an Interagency Task Force on Carbon Capture and Storage was established to develop a comprehensive and coordinated federal strategy to speed the commercial development and deployment of CCS technologies. The Task Force is specifically charged with proposing a plan to overcome the barriers to the widespread, cost-effective deployment of CCS within 10 years, with a goal of bringing 5 to 10 commercial demonstration projects online by 2016. As part of its work, the Task Force prepared a report that summarizes the state of CCS and identified technical and non-technical barriers to implementation. For additional information on the Task Force and its findings on CCS, go to: http://www.epa.gov/climatechange/policy task force.html. Because the development status of CCS technologies and their applicability to coal-fired EGUs are thoroughly discussed in the Task Force report, there will be no further discussion in this document.

3.2. Efficiency Improvements for Existing Coal-fired EGU Projects

Numerous efficiency improvements can be applied to coal-fired EGUs to increase thermal efficiency of power production (NETL, 2008, Sargent & Lundy, 2009; U.S. DOE, 2009; U.S. DOE, 2010). One specific example is the NETL study, which conducted a literature review of published articles and technical papers identifying potential efficiency improvement techniques applicable to existing coal-fired EGUs. Efficiency improvements can be expressed in different formats; they may be reported as an absolute change in overall efficiency (e.g., a change from 40% to 42% represents a 2% absolute increase). They may also be presented as the relative change in efficiency (e.g., a change from 40% to 42% is a relative change in efficiency and fuel use of 5%). The relative change in efficiency is the most consistent approach, since it corresponds to the same change in heat rate.

A summary of the findings from the NETL study is presented in **Exhibit 3-1**. The efficiency percentages were converted to a common basis so that all of the data could be compared. All of these improvements could not necessarily be implemented at each coal-fired EGU because of site-specific factors.

3.3. Efficiency Improvements for New Coal-Fired EGU Projects

3.3.1. Steam Cycle

The theoretical maximum achievable thermal efficiency achievable by an EGU using the Rankine cycle regardless of the technologies used is approximately 63% because of thermodynamic limitations and energy losses that cannot be recovered. Existing coal-fired EGUs using the Rankine cycle operate well below this limit. If the energy input to the cycle is kept constant, increasing the pressures and temperatures for the water-steam cycle will increase the output and the overall efficiency. However, a practical limitation to the higher pressure and temperatures that can be achieved in a boiler is the availability of boiler materials that can withstand these elevated conditions over an acceptable service life. The majority of existing PC-fired EGUs have subcritical boilers. Subcritical boilers typically operate at pressures of 2,400 psi (17 MPa) and at temperatures between 1,000 to 1,050°F (540 to 570°C). However, subcritical boilers can be designed to operate at steam pressures as high as 3,200 psi (22 MPa) and steam temperatures as high as 1,050°F (570°C).

The use of materials that can withstand the high-temperature and pressure of supercritical steam conditions allows for substantial improvements in efficiency for EGUs. "Supercritical" is a thermodynamic term describing the state of a substance where there is no clear distinction between the liquid and the gaseous phase (i.e., they are a homogenous fluid). Technically, the term "boiler" should not be used for a supercritical pressure steam generator, as no "boiling" actually occurs in this device, but it is common practice to use the term "Benson boiler." Supercritical EGUs typically use steam pressures of 3,500 psi (24 MPa) and steam temperatures of 1,075°F (580°C). However, supercritical boilers can be designed to operate at steam pressures as high as 3,600 psi (25 MPa) and steam temperatures as high as 1,100°F (590°C). Above this temperature and pressure the steam is sometimes called "ultra-supercritcal".

For a supercritical boiler, the feed water enters the boiler, is converted to steam, and is passed directly to the steam turbine (a supercritical boiler does not have a steam drum as shown in Exhibit 2-3 for a subcritical boiler PC-fired EGU configuration). Because the water-steam cycle medium is a single phase fluid with homogeneous properties, there is no need to separate steam from water in a drum. Supercritical boilers operate as once-through boilers in which the water and steam generated in the furnace waterwalls passes through only once. This eliminates the need for water/steam separation in drums during operation, and allows a simpler separator to be employed during start-up conditions. Because these units do not have thick-walled steam drums, their start-up times are quicker, further enhancing efficiency and plant economics. Due to the availability of steam turbines that are designed for supercritical steam conditions, supercritical applications are presently limited to facilities of approximately 200 MWe gross output or more. Supercritical boilers are a well-established technology, and over 500 supercritical plants are currently operating worldwide (VGB, 2008).

The majority of EGUs have a single reheat cycle where the steam is first passed through the high pressure portion of the steam turbine and is then reheated in the boiler prior to passing through the remainder of the turbine. This process increases the efficiency of the EGU without increasing the maximum steam temperature. An additional steam cycle improvement that further increases efficiency is the use of a double reheat cycle, which reduces fuel use by approximately 1.5% compared to a similar EGU using a single reheat cycle (Retzlaff, 1996). The efficiency benefits of using a double reheat cycle have been recognized since the 1960s. However, the additional cost of a double reheat cycle has made a single reheat cycle typical for the majority of EGUs in the U.S.

To establish performance and cost baselines for analyzing EGU technology, NETL funded an independent assessment of the cost and performance of fossil energy power systems. The assessment specifically includes PC-fired boilers, IGCC, and natural gas-fired combined cycle systems in a consistent technical and economic manner reflecting market conditions for plants starting operation in 2010 (NETL, 2007). Performance and cost estimates were prepared for each configuration with and without CO_2 CCS. The Total Plant Cost (TPC) and Operation and Maintenance (O&M) costs for each of the cases in the study were estimated in January 2007 dollars and assumed plant construction on a generic site. The costs do not include owner cost and additional costs for special site-specific considerations at a given site.

For the PC-fired systems, the NETL study included a comparison of subcritical (2,400 psig/1,050°F/1,050°F) and supercritical (3,500 psig/1,100°F/1,100°F) PC-fired EGUs, each rated at nominal 550 MWe net capacity and firing Illinois No. 6 bituminous coal. A summary comparing the results for the subcritical unit versus the supercritical unit (without CO_2 CCS) is presented in Exhibit 3-2. The analysis shows an efficiency (HHV) of 36.8% for the subcritical boiler compared to 39.1% for the supercritical boiler.

Exhibit 3-1. Existing coal-fired EGU efficiency improvements reported for actual efficiency improvement projects

Efficiency Improvement Technology	Description	Reported Efficiency Increase[a]
Combustion Control Optimization	Combustion controls adjust coal and air flow to optimize steam production for the steam turbine/generator set. However, combustion control for a coal-fired EGU is complex and impacts a number of important operating parameters including combustion efficiency, steam temperature, furnace slagging and fouling, and NOX formation. The technologies include instruments that measure carbon levels in ash, coal flow rates, air flow rates, CO levels, oxygen levels, slag deposits, and burner metrics as well as advanced coal nozzles and plasma assisted coal combustion.	0.15 to 0.84%
Cooling System Heat Loss Recovery	Recover a portion of the heat loss from the warm cooling water exiting the steam condenser prior to its circulation thorough a cooling tower or discharge to a water body. The identified technologies include replacing the cooling tower fill (heat transfer surface) and tuning the cooling tower and condenser.	0.2 to 1%
Flue Gas Heat Recovery	Flue gas exit temperature from the air preheater can range from 250 to 350°F depending on the acid dew point temperature of the flue gas, which is dependent on the concentration of vapor phase sulfuric acid and moisture. For power plants equipped with wet FGD systems, the flue gas is further cooled to approximately 125°F as it is sprayed with the FGD reagent slurry. However, it may be possible to recover some of this lost energy in the flue gas to preheat boiler feedwater via use of a condensing heat exchanger.	0.3 to 1.5%
Low-rank Coal Drying	Subbituminous and lignite coals contain relatively large amounts of moisture (15 to 40%) compared to bituminous coal (less than 10%). A significant amount of the heat released during combustion of low-rank coals is used to evaporate this moisture, rather than generate steam for the turbine. As a result, boiler efficiency is typically lower for plants burning low-rank coal. The technologies include using waste heat from the flue gas and/or cooling water systems to dry low-rank coal prior to combustion.	0.1 to 1.7%
Sootblower Optimization	Sootblowers intermittently inject high velocity jets of steam or air to clean coal ash deposits from boiler tube surfaces in order to maintain adequate heat transfer. Proper control of the timing and intensity of individual sootblowers is important to maintain steam temperature and boiler efficiency. The identified technologies include intelligent or neural-network sootblowing (i.e., sootblowing in response to real-time conditions in the boiler) and detonation sootblowing.	0.1 to 0.65%
Steam Turbine Design	There are recoverable energy losses that result from the mechanical design or physical condition of the steam turbine. For example, steam turbine manufacturers have improved the design of turbine blades and steam seals which can increase both efficiency and output (i.e., steam turbine dense pack technology).	0.84 to 2.6

Source: NETL, 2008

[a] Reported efficiency improvement metrics adjusted to common basis by conversion methodology assuming individual component efficiencies for a reference plant as follows: 87% boiler efficiency, 40% turbine efficiency, 98% generator efficiency, and 6% auxiliary load. Based on these assumptions, the reference power plant has an overall efficiency of 32% and a net heat rate of 10,600 Btu/kWh. As a result, if a particular efficiency improvement method was reported to achieve a 1% point increase in boiler efficiency, it would be converted to a 0.37 % point increase in overall efficiency. Likewise, a reported 100 Btu/kWh decrease in net heat rate would be converted to a 0.30% point increase in overall efficiency.

Exhibit 3-2. Summary of NETL performance, cost, and CO_2 emissions comparison analysis for nominal 550 MWe PC-fired EGU burning bituminous coal by steam cycle

Parameter	Bituminous Coal-Fired EGU	
	Subcritical Boiler	Supercritical Boiler
Gross Power Output (kWe)	583,315	580,260
Auxiliary Power Requirement (kWe)	32,780	30,110
Net Power Output (kWe)	550,445	550,150
Coal Flow Rate (lb/hr)	437,699	411,282
HHV Thermal Input (kW) Net Plant	1,496,479	1,406,161
HHV Net Efficiency (%)	36.8%	39.1%
Total Plant Cost ($ x 1,000)[a]	$852,612	$866,391
Total Plant Cost ($/kW)[a]	$1,549	1,575
Levelized Cost of Electricity (mills/kWh)[b]	64.0	63.3
CO_2 Emissions (ton/hr)	519.1	487.6
CO_2 Emissions (ton/year)	3,864,884	3,632,301
CO_2 Emissions (lb/MMBtu)	203	203
CO_2 Emissions (lb/MWh gross output)	1,780	1,681
CO_2 Emissions (lb/MWh net output)	1,886	1,773

Source: NETL, 2007. Exhibit ES-2.

[a] The NETL costs are presented as "overnight costs" in January 2007 dollars and do not include escalation, owner's costs, taxes, site specific considerations, labor incentives, etc.

[b] 10 mills are equivalent to 1 U.S. cent.

Continuing research and advances in metallurgy have allowed the development of supercritical boilers capable of operating at increasingly higher temperatures and pressures, achieving increasingly higher efficiencies. Ultra-supercritical (USC) boilers designed to operate at steam conditions in excess of 4,500 psi can potentially operate at efficiencies approaching 50% (HHV). Steam temperature and pressure selection for boilers depends in part upon fuel corrosiveness, and research is focused on the development of new materials for boiler tubes and high alloy steels that minimize corrosion. There are potential concerns that temperatures above 1,100°F (590°C) while firing high-sulfur coal (such as Illinois No. 6) would result in an exponential increase of the material degradation of the highest temperature portions of the superheater and reheater due to coal ash corrosion. This could require pressure parts replacement outages every 10 to 15 years. The availability and reliability of materials required to support the elevated temperature environment for high sulfur or chlorine applications, although extensively demonstrated in the laboratory, has not been fully demonstrated commercially (NETL, 2007). Additional factors that could limit steam temperatures and pressures are the maximum values specified in the American Society of Mechanical Engineers (ASME) Boiler and Pressure Vessel Code, Section I: Power Boilers. A developer deviating from this code could have difficulty acquiring insurance, or be out of compliance with specific state code requirements.

The commercial use of ultra-supercritical technology has historically been, and continues to be, prevalent in countries outside the U.S., such as Denmark, Germany, and Japan. Ultra-supercritical boilers burning various coal ranks are being widely deployed throughout the world. Construction of the first modern ultra-supercritical EGU built in the U.S. began in 2008 at the Southwest Electric Power Company's John W. Turk, Jr. Power Station near Texarkana, AR. This 600 MWe PC-fired facility will burn PRB subbituminous coal and is scheduled to begin operation in late 2012. Examples of ultra-supercritical PC-fired EGUs often cited as representing the currently highest efficiency operating coal-fired EGUs in the world include:

- 384 MWe ultra-supercritical PC-fired EGU for the European Vattenfall company's Nordjylland Power Station Unit 3 located near Aalborg, Denmark (Vattenfall, 2006). This power station began operation in 1998 and is a combined heat and power (CHP) facility that generates electricity and produces heat for a district heating system. The unit burns imported bituminous coal and uses seawater for cooling. Reported overall unit efficiency is 47% LHV or 45.3% HHV (International Energy Agency [IEA] Clean Coal Centre, 2007).
- 965 MWe ultra-supercritical PC-fired EGU for the German RWE Power company's Niederaussem Power Station Unit K located near Cologne, Germany (RWE Power, 2004). This unit burns lignite with a 51 to 58% moisture content and started operation in 2002. Reported unit operating efficiency is 43.2% LHV or 37% HHV (IEA Clean Coal Centre, 2007).
- Two 600 MWe ultra-supercritical PC-fired EGUs for the J-POWER (Electric Power Development Co., Ltd.) Isogo Thermal Power Station New Unit 1 and New Unit 2 located in Yokohama, Japan (J-POWER, 2009). These units burn domestic and imported bituminous coal and use seawater for cooling. New Unit 1 started operation in 2002 and has a reported unit operating efficiency of 42% LHV or 40.6% HHV (IEA Clean Coal Centre, 2007). New Unit 2 started commercial operation in July 2009. The New No. 2 Plant, has a higher efficiency due to the boosting the reheat steam temperature 18°F higher than the New Unit No. 1 to 1,148°F (J-POWER, 2009).
- 450 MWe ultra-supercritical PC-fired EGU for the Capital Power Corporation's Genesee Power Station Unit 3 located near Edmonton, Alberta, Canada (Peltier, 2005). This unit burns Alberta subbituminous coal. This unit began operation in 2005 and has a reported unit operating net efficiency or 41% LHV or 39.6% HHV (IEA Clean Coal Centre, 2007).

As part of NETL's performance and cost baseline analysis for electricity production (NETL, 2007), IGCC EGUs were analyzed for the General Electric Energy (GEE), ConocoPhillips (CoP), and Shell coal gasification processes. The IGCC cases have different gross and net power outputs than the PC-fired EGU cases because of the combustion turbine size constraint. The advanced F-class turbine used to model the IGCC cases comes in a standard size of 232 MWe when operated on syngas. Each IGCC case uses two combustion turbines for a combined gross output of 464 MWe. Additional electrical output is generated by steam turbines, with steam from the HRSGs extracting heat from the combustion turbine exhaust. Although the two combustion turbines provide 464 MWe gross output in all cases, the overall combined cycle gross output ranges from 742 to 770 MWe. The net outputs range from 623 to 640 MWe depending on the gasification process. A summary comparison of the results for the three gasification processes without CO_2 CCS is presented in Exhibit 3-3. Although the efficiency of the combined cycle block is approximately 50% efficient in converting the syngas to electricity, parasitic loads of the gasification process lower the net efficiency for the IGCC to 38.2 to 41.1% HHV.

Exhibit 3-3. Summary of NETL performance, cost, and CO_2 emissions comparison for an IGCC power plant by gasification process

Parameter	Gasification Process		
	GEE Radiant	CoP E-GasTM	Shell
Gross Power Output (kWe)	770,350	742,510	748,020
Auxiliary Power Requirement (kWe)	130,100	119,140	112,170
Net Power Output (kWe)	640,250	623,370	635,850
Coal Flow Rate (lb/hr)	489,634	463,889	452,620
HHV Thermal Input (kW) Net Plant	1,674,044	1,586,023	1,547,493
HHV Efficiency (%)	38.2%	39.3%	41.1%
Total Plant Cost ($ x 1,000)[a]	1,160,919	1,078,166	1,547,483
Total Plant Cost ($/kW)[a]	1,813	1,733	1,977
Levelized Cost of Electricity (mills/kWh)[b]	78.0	75.3	80.5
CO_2 Emissions (ton/hr)	561.9	539.1	527.1
CO_2 Emissions (ton/year)	3,937,728	3,777,815	3,693,990
CO_2 Emissions (lb/MMBtu)	197	199	200
CO_2 Emissions (lb/MWh gross output)	1,459	1,452	1,409
CO_2 Emissions (lb/MWh net output)	1,755	1,730	1,658

Source: NETL, 2007. Exhibit ES-2.

[a] The NETL costs are presented as "overnight costs" in January 2007 dollars do not include escalation, owner's costs, taxes, site specific considerations, labor incentives, etc.

[b] 10 mills are equivalent to 1 U.S. cent.

3.3.2. Coal Drying

Low-rank coals (lignite and subbituminous) are often utilized because the low cost per unit of heat input relative to bituminous coal and the low sulfur content. However, a major disadvantage of low-rank coals is their high moisture content, typically 25 to 40%. When this coal is burned, considerable energy is required to vaporize and heat the moisture, thus raising the heat rate of the EGU and lowering its efficiency. As fuel moisture decreases, the heating value of the fuel increases so that less coal needs to be fired to produce the same amount of electric power. Drier coal is also easier to handle, convey, and pulverize – reducing the burden on the coal-handling system. In addition, an EGU boiler designed for dried coal is smaller and has lower capital costs than a comparable EGU designed to burn coal that has not been dried. The pre-combustion drying of low-rank coals can improve the overall efficiency and several advanced coal drying technologies are or nearly are commercial available.

Great River Energy developed a coal drying technology for low-rank coals in partnership with the U.S. DOE as part of the DOE's Clean Coal Power Initiative (U.S. DOE, 2007). The technology has been successfully demonstrated on a PC-fired boiler burning lignite at the utility's Coal Creek Station in Underwood, ND. The technology is now commercially offered under the trade name DryFiningTM (U.S. DOE, 2010b). The DryFiningTM process passes warm cooling water from the steam turbine exhaust condenser through an air heater where ambient air is heated before being sent to a fluidized bed coal dryer. The dried coal leaving the fluidized bed is sent to a pulverizer and then to the boiler. Air leaving the fluidized bed is filtered before being vented to the atmosphere. In addition to using power plant waste heat to reduce moisture, DryFiningTM also segregates particles by density. This means a significant amount of higher density compounds containing sulfur and mercury can be sorted out and returned to the mine rather than utilized in the boiler. The end result is that more energy can be extracted from the coal while simultaneously reducing emissions of mercury, sulfur dioxide, and NOX. At the Coal Creek Station, the process increased the energy content of the lignite from 6,200 to 7,100 Btu/lb, thereby resulting in a decrease in the fuel input into the boilers by 4% and a corresponding decrease in CO_2 emissions gains in overall efficiency of 2 to 4% are reported for the process.

RWE Power in Germany is also developing a fluidized bed drying technology for lignite, called WTA (RWE Power, 2009). A fundamental difference between the two drying processes is the WTA process first mills then dries the lignite while the DryFining^{TM} process first dries then mills the lignite. A prototype commercial-scale drying plant using the WTA process began operation in 2009 at the utility's Nederaussem Power Station site. For the WTA process, lignite is first milled to a fine particle size by hammer mills in direct series with a two-stage fluidized-bed dryer. The dried fuel exiting the stationary bed is separated from the gas stream and mixed with coarser lignite solids collected from the bottom of the dryer bed and then fed directly to the boiler. The heat needed for the drying of the fuel is supplied by external steam, which is normally taken from the turbine with the heat transfer taking place in tube bundles located inside the bed. Based on the development work completed to date of the WTA technology, the net gain in cycle efficiency is reported to be on the order of 4 percentage points, depending on the moisture content of the raw coal and the final moisture of the dried lignite.

Several other coal drying technologies are in ongoing development. One coal drying process being developed by DBAGlobal Australia Pty, Ltd., with the trade name Drycol process uses the controlled application of microwave radiation to dry coal (Graham, 2007). Coal feed stock is first separated into fine grade coal and one or more larger grades. The fine coal is loaded onto a conveyor and conveyed continuously through a microwave-energized heating chamber for drying. The fine grade coal is dried sufficiently so that when it is recombined with the larger grade coals, the moisture content of the aggregate coal is within a target moisture content range. Other coal drying technologies for low-rank coals in various stages of development include: 1) attrition milling of coal followed by air drying to produce a low-moisture coal product, 2) compressing heated, coarse crushed coal to squeeze water out , and 3) heating wet coal under pressure to approximately 480 to 570°F (APP, 2008).

To date, it has not been economic to dry subbituminous coal at the mine prior to transport to an EGU. In addition, subbituminous coal that has been dried results in increased coal dust, can spontaneously combust, and will reabsorb moisture during transport and storage. However, the development of more efficient drying technologies such as dryers using flue gas recirculation and briquetting of the dried coal to avoid spontaneous combustion and moisture reabsorption can improve the economics of upgrading low-rank coals. Descriptions of upgrading low rank coals are available from the White Energy Company (http://www.white

energyco.com) and Evergreen Energy (http://www.evgenergy.com/). From an overall GHG perspective, the increased EGU efficiency and decreased transportation GHG emissions would have to be compared to the energy required to dry and process the coal at the mine.

3.3.3. Boiler Feedwater Heating & Hot-Windbox

The high-pressure liquid water entering the steam generator is called feedwater. A feedwater heater is an EGU component used to pre-heat water delivered to the boiler section. Thermodynamic optimization of this cycle is important to overall EGU efficiency. In a conventional EGU, the energy used to heat the feedwater is steam extracted between the stages of the steam turbine (see Exhibit 2-3). Therefore, approximately a quarter of the steam that would be used to perform expansion work in the turbine (and generate power) is not utilized for that purpose. However, using other heat sources for the feedwater heater avoids the need to extract steam from the turbine allowing the steam to be used for electric power generation and increases the output of the steam cycle and potentially lowers GHG emissions. This alternate heat source can either be from an integrated solar thermal energy source or from a combustion turbine. Examples of solar thermal energy used to augment the steam cycle at combined cycle facilities include the Martin Next Generation Solar Center in Florida and the proposed Green Energy Partners/Stonewall, LLC facility in Virginia. The first coal-fired power plant to integrate solar thermal technology is the Cameo generating station in Colorado. In addition, EPRI (Electric Power Research Institute) is currently evaluating adding solar thermal energy to the Escalante and Mayo coal-fired power plants. An example of combustion turbine integration for feedwater heating is the Kettle Falls Generating Station (Schimmoller, 2003). For coal-fired boiler systems optimized to accommodate the combustion turbine exhaust, the incremental fuel efficiencies would be expected be comparable with combined cycle generation (Escosa, 2009; Stenzel). Another potential approach to integrate the use of a combustion turbine with a coal-fired steam cycle is using the turbine exhaust directly in the boiler in a hot-windbox. This involves injection of the combustion turbine exhaust directly into the boiler windbox or primary air ducts to provide an oxygen source as well as a heat source.

3.4. Combined Heat and Power Plant

Coal-fired EGUs dedicated to electric power generation and using the latest commercially available advanced technologies will generally operate at overall net efficiencies of approximately 40%. Significant amounts of energy released by coal combustion are lost during the steam condensation segment of the Rankine cycle due to heat transfer into the cooling water. In Europe, electricity is commonly generated by facilities that serve as both electricity generators and thermal energy producers for the local town or city district heating system. These combined heat and power (CHP) facilities are also known as cogeneration facilities. Operating an electric power station in a CHP mode allows recovery of some of the heat that would otherwise be rejected into cooling water, improving the overall efficiency of energy utilization. In applying CHP to an existing or new EGU, the temperature of the cooling water is normally not high enough to meet the requirements for most district heating or industrial process applications. In these cases, steam would be extracted at an elevated pressure and temperature from an intermediate stage of the steam turbine and then used for district or process heating.

This results in a decrease in the total electric power generation from the EGU. However, the overall fuel efficiency of CHP is higher than if electricity and steam were generated separately.

Because electricity can be transmitted over long distances, electric power plants can be located in remote areas as well as urban areas. However, thermal energy cannot be effectively transported over extended distances. This limits the practicality of incorporating a CHP mode into many electric power plant designs. The EGU needs to be located in close proximity to either a district energy system or an industrial facility with a significant and steady thermal demand. There are a number of examples; however, where industrial facilities have collocated with existing or new coal power plants in order to have access to reliable, low cost steam:

- DuPont's titanium dioxide plant in Johnsonville, Tennessee, is located next to TVA's Johnsonville power plant and buys high pressure process steam from the 1,200 MW facility. The power plant is comprised of 10 coal-fired boiler steam turbine units; the DuPont plant uses steam extracted from Units 1 through 4. Providing steam to the DuPont facility at the required process pressures reduces overall output of the power plant by 50 MW.
- Blue Flint Ethanol in Underwood, North Dakota, is a 50 million gallon per year dry mill ethanol producer located next to Great River Energy's 1,160 MW Coal Creek lignite-fired power plant. Starting operations in 2007, the ethanol facility purchases approximately 100,000 pounds per hour of medium pressure steam extracted from the power plant.
- Goodland Energy Center in Goodland, Kansas, is a small 22 MW coal-fired power plant that is supplying steam to a 20 million gallon per year ethanol plant and a 12 million gallon per year biodiesel plant, both co-located with the power plant. The power plant and the ethanol plant both started construction in 2006.

3.5. Oxygen Combustion

Oxygen combustion (oxy-combustion, oxy-firing or oxy-fuel) is an emerging technology applicable to either new or existing EGUs. The advantage offered by this technology is its potential for CO_2 emissions control because it produces a concentrated (nearly pure) CO_2 exhaust gas stream that requires minimal post-combustion clean-up prior to compression, transportation, and injection for long term storage. The basic concept of oxy-combustion is to use a mixture of oxygen (or oxygen-enriched air) and recycled flue gas (containing mostly CO_2) in place of ambient air for coal combustion. The resulting flue gas contains primarily CO_2 and water vapor with smaller amounts of oxygen, nitrogen, s_{O2}, and NOX. Consequently, the flue gas can be processed relatively easily to further purify the CO_2 (if necessary) for use in enhanced oil or gas recovery or for geological storage.

An oxy-combustion power plant consists of an air separation unit (ASU), an EGU with O_2-blown combustion, and a CO_2 treatment unit. The conventional ASU is a cryogenic process that has a significant energy requirement. However, alternative oxygen separation methods are being researched for possible commercial scale development. These alternative methods include ion transport membranes (ITM), ceramic autothermal recovery, oxygen transport membranes, and chemical looping (UARG, 2008). Oxygen is mixed with recirculated flue gas to create a mixture of O_2 and CO_2 (and some H_2O) which is used as the source of combustion oxidant instead of ambient air. The absence of air nitrogen produces a flue gas stream with a high concentration of CO_2.

Several research institutes are focusing on laboratory- and pilot-scale testing of oxy-fuel combustion (Levasseur, 2009). Pilot test programs currently are being conducted for European Enhanced Capture of CO_2 (ENCAP) program and the Advanced Development of the Coal-Fired Oxyfuel Process with CO_2 Separation (ADECOS) program. Additional research and development programs are being conducted, including:

- A 30 MW oxy-firing pilot plant at the Schwarze Pumpe station in Spremberg, Germany. This plant is the first complete oxy-combustion unit that includes the integrated system from the air separation unit to the gas purification and compression systems. The CO_2 will be compressed and liquefied for storage experiments to be conducted.
- A 32 MW oxy-firing demonstration project in France retrofitting an existing boiler to natural gas oxy-combustion. The captured CO_2 will be transported through an approximately 19 mile long pipeline and stored in a depleted gas field in Lacq, South of France.
- A comprehensive test program using the 15 MW tangentially-fired Boiler Simulation Facility and 15 MW Industrial Scale Test Facility operated by Alstom Power, Inc., in Windsor, CT. Testing is being conducted to assess a broad range of oxy-combustion design options. Project partners include the U.S. DOE, the Illinois Clean Coal Institute, and 10 electric utility companies.

4. COAL-FIRED EGU TECHNOLOGY ALTERNATIVES ANALYSIS

There is no one best available coal-fired EGU technology universally applicable to all EGU projects. The coal-fired EGU technology alternatives most suitable for a given project must be evaluated on a site-specific basis. An evaluation for a new facility would include the use of carbon capture and storage and the most efficient technologies (e.g., ultra-supercritical steam conditions, IGCC, pressurized fluidized bed), double steam reheat, coal drying, FGD technology, and CHP.

4.1. Site-Specific Coal-Fired EGU Technology Alternatives Analysis Example

This section summarizes the results for analyses prepared in support of an air permit application for a new 830 MW supercritical PC-fired EGU to be built by the Consumers Energy Company at their Karn-Weadock Generating Station in Bay County, Michigan. The EGU is designed to burn PRB subbituminous coal, but can also mix bituminous coal based on supply or price variations. A supercritical PC boiler with steam turbine throttle pressure of

3,805 psia and superheat and single reheat temperatures of 1,100 °F was proposed by the Consumers Energy Company.

The air permit application was initially submitted to the State of Michigan Department of Natural Resources and Environment (DNRE) in 2007. As an amendment to the permit application for the Consumers Energy project, the DNRE requested that Consumers Energy Company prepare and submit a top-down BACT analysis for IGCC plant as an alternative to building the supercritical PC-fired EGU. The supercritical PC-fired EGU air emissions controls included selective catalytic reduction (SCR), baghouse, wet flue gas desulfurization, hydrated lime injection, and activated carbon injection. The IGCC plant air emissions controls included SCR, Selexol or Sulfinol, black water handling equipment, and sulfur-impregnated activated carbon.

To estimate the IGCC plant capital costs, the permit applicant developed a premium factor for constructing an IGCC plant compared to a similar capacity supercritical PC-fired EGU based on information collected by the permit applicant including reported costs of other IGCC projects. This capital cost premium factor was then applied to the estimated costs for the supercritical PC-fired EGU. Additional cost estimates were made for fuel, water consumption, waste disposal, operation and maintenance, and pollutant allowance purchases required for each facility type. The reference IGCC plant capital costs were estimated to be approximately 24% higher than the supercritical PC-fired EGU. The analysis estimated that the cost of electricity generation from an IGCC unit would be approximately 37% higher for the IGCC unit than for the supercritical PC-fired EGU. The projected cost of generation for the supercritical PC-fired EGU was $60/MWh compared to $95/MWh for the IGCC plant. The cost estimates prepared for the analysis are summarized in Exhibit 4-1.

Exhibit 4-1. Supercritical PC-fired EGU and IGCC plant cost comparison Summary prepared for Consumers Energy EGU project

Cost Parameter[a]	800 MWe net Supercritical PC-fired EGU	800 MWe net IGCC Plant	Difference (IGCC vs. PC)
Capital Costs ($)	$2,671,916,111	$3,526,039,934	$854,123,823
Annualized Costs ($/yr)	$273,871,401	$361,419,093	$87,547,692
Fuel Cost ($/yr)	$117,624,738	$130,679,946	$13,055,208
Cooling Water Consumption Cost ($/yr)	$5,166,656	$3,114,268	(-$2,052,388)
Waste Disposal Cost ($/yr)	$316,937	$201,042	(-$115,895)
Operating and Maintenance Costs ($/yr)	$59,307,369	$84,336,386	$25,029,017
Total Annual Cost ($/yr)	$456,287,101	$579,750,735	$123,463,634
Annualized Cost of SO_2 Allowances ($/yr)	$2,442,975	$610,531	(-$1,832,445)
Annualized Cost of NO_x Allowances ($/yr)	$4,132,417	$2,611,787	(-$1,520,630)

Source: Consumers Energy Company, 2008.

[a] Costs are presented in September 2007 dollars and include owner's and financing costs.

Exhibit 4-2. Coal-fired EGU technology alternatives cost comparison summary prepared for Consumers Energy EGU project

Coal-Fired EGU Technology	40-yr. BusBar Cost Excluding CO_2 Cost[a]	40-yr. BusBar Cost Including CO_2 Cost[a]	Availability	Efficiency- Heat Rate (Btu/kWh)	CO_2 Emission Rate (tons/MWh)
Supercritical PC-fired 830 MW	$97 per MWh	$133 per MWh	86% to 92%	9,134	0.94
Subcritical PC-fired	$101 per MWh	$136 per MWh	84% to 89%	9,407	0.97
Subcritical CFB boiler	$108 per MWh	$145 per MWh	87%	9,798	1.01
Supercritical CFB boiler	$108 per MWh	$144 per MWh	87%	9,508	0.98
IGCC plant	$128 per MWh	$162 per MWh	70% to 81%	9,490	0.93
Ultra-supercritical PC-fired	$98 per MWh	$133 per MWh	91%	9,019	0.93
Supercritical PC-fired with Carbon Capture and Storage (CCS)	$135 per MWh	$139 per MWh	86% to 91%	10,836	0.11
Supercritical PC-fired 500 MW	$104 per MWh	$140 per MWh	86% to 92%	9,134	

Source: Consumers Energy Company, 2009.

[a] Busbar costs is the cost to generate the power leaving the plant (beyond the generator but prior to the voltage transformation point in the plant switchyard) and include all plant fixed costs (including all costs associated with the capital investment), fuel costs, operating and maintenance costs, emissions costs, interconnection costs, and transmission system upgrade costs. A busbar cost excluding CO_2 costs assumes a CO_2 tax or cap-and-trade program has not been implemented. A busbar cost including CO_2 costs assumes a CO_2 tax cost of $22/ton beginning in 2012 and rising to $53/ton by 2025.

At the request of the State air permitting authority for the project, the permit applicant prepared and submitted a second electric generation alternatives analysis for the Consumers Energy project in June 2009 (Consumers Energy Company, 2009). As part of this alternatives analysis, a comparison of the various coal-fired EGU technologies was presented with respect to air emissions from coal combustion including CO_2. The ultra-supercritical PC-fired EGU has the lowest projected heat rate of the analyzed technologies. Of the technologies without carbon capture and storage, it also has the lowest GHG emissions rate. In December 2009, the Michigan Department of Environmental Quality issued the construction permit based on the proposed supercritical PC boiler. The results are summarized in Exhibit 4-2.

4.2. EPA GHG Mitigation Database

The EPA Office of Research & Development (ORD) is collecting information regarding CO_2 mitigation measures applicable to coal-fired EGUs for compilation in a publicly-accessible GHG Mitigation Database. Version 1 of this database is expected to be released to the public in late summer 2010. The database is a tool that provides information of both commercially available technologies, as well as emerging technologies that are being demonstrated at larger scales for commercial viability.

REFERENCES

Internet Web Site addresses cited for individual references were available as of the date of publication of this document.

Asai, Akihisu, et al. 2004. *System Outline and Operational Status of Karita Power Station New Unit 1 (PFBC)*. JSME International Journal, Series B. Vol. 47, No. 2, 2004. Pp. 193-199. Available at: <http://www.jstage.jst.go.jp/article/jsmeb/47/2/193/_pdf>.

Asia-Pacific Partnership on Clean Development and Climate (APP). 2008. *Brown Coal Drying Technologies*. APP Joint Meeting of Cleaner Fossil Energy and Power Generation and Transmission Task Forces. April 1, 2008. Available at: <http://www.asiapacificpartnership.org/pdf/CFE/meeting_melbourne/SurveyofBrownCoalDr yingTechnologies-Godfrey.pdf>.

Benson, Lewis, et al. 2003. *Control of Sulfur Dioxide and Sulfur Trioxide Using By-Product of a Magnesium-Enhanced Lime FGD System*. Presented at ICAC Forum '03 Multi-Pollutant Emission Controls & Strategies, Multi-Pollutant Emission Controls & Strategies, Nashville, TN . October 14-15, 2003. Available at: <http://www.carmeusena.com/files/files/techpapersreports/carmeuse_icac_20forum_2003 .pd f>.

Benson, Lewis. *New Magnesium-Enhanced Lime Flue Gas Desulfurization Process*. Technical Paper Report, Carmeuse North America. Pittsburgh, PA. Available at: <http://www.carmeusena.com/files/files/TechPapersReports/fgd_new_magnesium.pdf >.

Consumers Energy Company. 2008. *Amendment to Application No. 341-07 Consumers Energy Company – Bay Count Top-Down BACT Analysis for IGCC*. April 2008. Consumers Energy Company, Jackson, MI. Available at:

<http://www.deq.state.mi.us/aps/downloads/permits/CFPP/2007/341-07/Top-Down%20BACT%20Analysis%20for%20IGCC.pdf >.

Consumers Energy Company. 2009. *Balanced Energy Initiative: Electric Generation Alternatives Analysis.* June 2009. Consumers Energy Company, Jackson, MI. Available at: <http://www.deq.state.mi.us/aps/downloads/permits/PubNotice/341- 07/Alternatives Analysis.pdf >.

Davenport, W.G., et al. 2006. *Sulfuric Acid Manufacture.* Southern African Pyrometallurgy. South African Institute of Mining and Metallurgy, Johannesburg, 5-8 March 5-8, 2006.

Escosa, Jesús M. and Luis M. Romeo. 2009. *Optimizing CO$_2$ Avoided Cost by Means of Repowering.* Applied Energy, 86 (2009) 2351–2358.

Foster Wheeler North America Corp. 2009. *Utility CFB Goes Supercritical - Foster Wheeler's Lagisza 460 MWe Operating Experience and New 600 - 800 MWe Designs.* Prepared by James Utt, Arto Hotta, and Stephen Goidich, Foster Wheeler North America Corp., Clinton, NJ, for presentation at Coal-Gen 2009, Charlotte, NC, August 19-21, 2009. Available at:

<http://www.fwc.com/publications/tech_papers/files/TP_CFB_09_12.pdf>.

Graham, James. 2007. *Microwaves for Coal Quality Improvement: The Drycol Project.* DBAGlobal Australia, Milton Queensland, Australia. Presented at the SACPS/International Pittsburgh Coal Conference 2007, Johannesburg, South Africa, September 10-14, 2007.

Available at: <http://www.drycol.com/downloads/Drycol%20Paper%20ACPS1%20060608. pdf>.

He, Boshu, et al. 2002. *Temperature Impact on SO$_2$ Removal Efficiency by Ammonia Gas Scrubbing*, Energy Conversion and Management 44:2175-2188.
http://www.sciencedirect.com/science?_ob=ArticleURL&_udi=B6V02-3WRC6DR3&_user=775537&_origUdi=B6V2P-47CY482-G&_fmt=high&_coverDate=07%2F31%2F1998&_rdoc=1&_orig=article&_acct=C0000429
38&_version=1&_urlVersion=0&_userid=775537&md5=e2354e5058cba8f383722059ff
ecd1 cd

Hong, B.D. and E.R. Slatick. 1994. *Carbon Dioxide Emission Factors for Coal.* Originally published in U.S. Energy Information Administration, Quarterly Coal Report, January-April 1994, DOE/EIA-0121(94/Q1). Washington, DC, August 1994, pp. 1-8. Available at: <http://www.eia.doe.gov/cneaf/coal/quarterly/co$_2$_article/co2.html>.

IEA Clean Coal Centre. 2007. *G8 Case Studies by the IEA Clean Coal Centre.* Prepared by Colin Henderson, IEA Clean Coal Centre, London, United Kingdom for presentation at Third International Conference on Clean Coal Technologies for Our Future, Sotacarbo Coal Research Centre, Carbonia Sardinia, Italy. May 15 -17, 2007. Available at:<http://www.iea-coal.org/publishor/system/component_view.asp?LogDocId=81704&PhyDocId=6360>.

J-POWER (Electric Power Development Co., Ltd.). 2009. *Annual Report 2009.* Electric Power Development Co., Ltd., Tokyo, Japan. October 2009. Available at: <http://www.jpower.co.jp/english/ir/pdf/2009.pdf>.

Kaplan, P. Ozge, et al. 2008. *Is It Better To Burn or Bury Waste for Clean Electricity Generation?.* Environmental Science & Technology, Vol. 43, No. 6, 2009, pp. 1711-1717.

Korhonen, S., et al. 2001. *Methane and Nitrous Oxide Emissions in the Finnish Energy Production*. Fortum Tech-4615.

Levasseur, A. A., et al.. 2009. *Alstom's Oxy-Firing Technology Development and Demonstration - Near Term CO_2 Solutions*. Presented at the 34th International Technical Conference on Clean Coal & Fuel Systems, Clearwater, FL. May 31 - June 4, 2009. Available   at: <http://www.netl.doe.gov/technologies/coalpower/ewr/co$_2$/pubs/5290%20Alstom%20oxy combustion%20paper%20Clearwater%20jun09.pdf>.

Maziuk, John and John Kumm. 2002. *Comparison of Dry Injection Acid-Gas Control Technologies*. Presented at the 95th Annual Conference and Exhibition of the Air and Waste Management Association, Baltimore, MD. June 23-27, 2002.

Michigan Department of Natural Resources and Environment. 2009. PTI Application No. 341-07. December 29, 2009. Available at: <http://www.deq.state.mi.us/aps/downloads/permits/CFPP/2007/341-07/341-07.htm>.

National Energy Technology Laboratory (NETL), 2007. *Cost and Performance Baseline for Fossil Energy Plants, Volume 1: Bituminous Coal and Natural Gas to Electricity, Revision 1*. DOE/NETL-2007/1281. U.S. Department of Energy, National Energy Technology Laboratory, Pittsburgh, PA. August 2007. Available at: <http://www.netl.doe.gov/energyanalyses/pubs/Bituminous%20Baseline_Final%20 Report.pdf>,

National Energy Technology Laboratory (NETL), 2008. *Reducing CO_2 Emissions by Improving the Efficiency of the Existing Coal-fired Power Plant Fleet*, DOE/NETL-2008/1329. U.S. Department of Energy, National Energy Technology Laboratory, Pittsburgh, PA. July 23, 2008. Available at: <http://www.netl.doe.gov/energy-analyses/pubs/CFPP%20EfficiencyFINAL.pdf>.

National Energy Technology Laboratory (NETL), 2010a. *Overview of DOE's Gasification Program*. Presentation by Jenny Tennant, Technology Manager, Gasification, U.S. U.S. Department of Energy, National Energy Technology Laboratory, Pittsburgh, PA. January 25, 2010. Available at: <http://www.netl.doe.gov/technologies/coalpower/gasification/pubs/pdf/DOE%20Gasific atio n%20Program%20Overview%202010%2001-25%20v1v.pdf>.

National Energy Technology Laboratory (NETL), 2010b. *CCPI/Clean Coal Demonstrations Nucla CFB Demonstration Project, Project Fact Sheet*. U.S. Department of Energy, National Energy Technology Laboratory, Pittsburgh, PA. Accessed June 21, 2010. Available at: <http://www.netl.doe.gov/technologies/coalpower/cctc/summaries/ nucla/nuclademo.html>.

National Energy Technology Laboratory (NETL), 2010c. *CCPI/Clean Coal Demonstrations Tidd PFBC Demonstration Project, Project Fact Sheet*. U.S. Department of Energy, National Energy Technology Laboratory, Pittsburgh, PA. Accessed June 21, 2010. Available at: <http://www.google.com/imgres?imgurl=http://www.netl.doe.gov/ technologies/coalpower/cc tc/summaries/tidd/images/tidd_plant_bw.jpg&imgrefurl= http://www.netl.doe.gov/technologies/coalpower/cctc/summaries/tidd/ tidddemo.html&usg=__PMP6wAdGSJtmGwsCiNKKCwryUc=&h=>

Pacific Northwest National Laboratory (PNNL), 2009. *An Assessment of the Commercial Availability of Carbon Dioxide Capture and Storage Technologies as of June 2009*,

PNNL18520. Pacific Northwest National Laboratory, Richland, WA. June 2009. http://www.pnl.gov/main/publications/external/technical_reports/PNNL-18520.pdf>.

Peltier, Robert. 2005. *Genesee Phase 3, Edmonton, Alberta, Canada*. Power. July/August 2005. Available at: <http://www.epcor.ca/SiteCollectionDocuments/Corporate/pdfs/publications%20and%20ne wsletters/plattspower05.pdf>.

Peltier, Robert. 2010. *Plant Efficiency: Begin with the Right Definitions*. Power. February 1, 2010. Available at: <http://www.powermag.com/gas/Plant-Efficiency-Begin-with-the-RightDefinitions_2432.html>.

Retzlaff, Klaus M and W. Anthony Ruegger, 1996. *Steam Turbines for Ultrasupercritical Power Plants*. GE Power Generation, GER-3945A. Available at: <http://www.gepower.com/prod serv/products/tech docs/en/downloads/ger3945a.pdf>.

RWE Power. 2004. *Niederaussem Power Plant: A Plant Full Of Energy*. RWE Power Aktiengesellschaf, Essen, Germany. October 2004. Available at: <http://www.debriv.de/tools/download.php?filedata=1218532199.pdf&filename=Kraftw erk %20Niederaussem%20(englisch).pdf&mimetype=application/pdf>.

RWE Power. 2009. *WTA Technology: A Modern Process for Treating and Drying Lignite*. RWE Power Aktiengesellschaf, Essen, Germany. February 2009. Available at: <http://www.rwe.com/web/cms/mediablob/en/88166/data/183490/36906/rwe/innovations /po wer-generation/coal-innovation-centre/fluidized-bed-drying/download-wta-en.pdf>.

Sargent & Lundy. 2009. *Coal-Fired Power Plant Heat Rate Reductions*. Report Number SL-009597. Sargent & Lundy, Chicago, Il. January 22, 2009. Available at: <http://www.epa.gov/airmarkt/resource/docs/coalfired.pdf>.

Schimmoller, Brian, 2003. *Avista Kettle Falls*. Power Engineering. October, 2003. Available at: <http://www.powergenworldwide.com/ index/display/articledisplay/189185/ articles/ powerengineering/volume-107/issue-10/departments/managing-the-plant/avista-kettle-falls.html>

Shand, Mark A., 2009. *A New Look at SO_2 Removal*. April 28, 2009. Available at: <http://www.risiinfo.com/technologyarchives/chemicalsnews>.

Srivastava, Ravi K, Wojciech Jozewicz and Carl Singer, 2001. *SO2 Scrubbing Technologies: A Review*. Environmental Progress, 20(4):219-28.

Srivastava, Ravi K and Wojciech Jozewicz. 2001. *Flue Gas Desulfurization: The State of the Art*, Journal of the Air & Waste Management Association, 51:1676-88.

Stenzel, William C., et al. *Repowering Existing Fossil Steam Plants*. December 11, 1997. Available at: <http://soapp.epri.com/press/default.htm>.

Utility Air Regulatory Group (UARG). 2008. *A Review of Carbon Capture and Sequestration (CCS) Technology*, Prepared by J. Edward Cichanowicz, Saratoga, CA, December 2008.

U.S. Department of Energy (DOE). 1996. *Electric Utility Engineer's FGD Manual, Vol. 1--FGD Process Design*. U.S. Department of Energy, Office of Fossil Energy, Morgantown, WV. March 1996. Available at: < http://www.netl.doe.gov/technologies/coalpower/ewr/pubs/fgdmanual_vol1.pdf>.

U.S. Department of Energy (DOE). 2007. *Clean Coal Technology Power Plant Optimization Demonstration Projects Topical Report Number 25*. September 2007. Available at: <http://www.netl.doe.gov/technologies/coalpower/cctc/topicalreports/pdfs/topical25.pdf>

U.S. Department of Energy (DOE), 2009. *Technical Workshop Report: Opportunities to Improve the Efficiency of Existing Coal-Fired Power Plants*. Workshop sponsored by the U.S. Department of Energy and National Energy Technology Laboratory. Rosemont, IL.
July 15-16, 2009. Available at: <http://www.netl.doe.gov/energy-analyses/pubs/NETL%20Power%20Plant%20Efficiency%20Workshop%20Report%20Final. pdf>.

U.S. Department of Energy (DOE), 2010. *Technical Workshop Report: Improving the Efficiency of Coal-fired Power Plants in the United States*. Workshop sponsored by the U.S. Department of Energy and National Energy Technology Laboratory. Baltimore, MD. February 24-25, 2010. Available at: <http://www.netl.doe.gov/energy-analyses/refshelf/detail.asp?pubID=306>.

U.S. Department of Energy (DOE). 2010b. *Innovative Drying Technology Extracts More Energy from High Moisture Coal*. Fossil Energy Techline. March 11, 2010. Available at: <http://fossil.energy.gov/news/techlines/2010/10006-CCPI Technology Goes Commercial.html>.

U.S. Energy Information Administration (U.S. EIA). 2008. *EIA-923 (Schedule 2) - Monthly Utility and Nonutility Fuel Receipts and Fuel Quality Data (2008 Final Data)*. U.S. Energy Information Administration, Office of Coal, Nuclear, Electric and Alternate Fuels, U.S. Department of Energy, Washington, DC 20585. Accessed June 21, 2010. Available at: <http://www.eia.doe.gov/cneaf/electricity/page/eia423.html>.

U.S. Energy Information Administration (U.S. EIA). 2010. *Electric Power Annual 2008*. DOE/EIA-0348(2008). U.S. Energy Information Administration, Office of Coal, Nuclear, Electric and Alternate Fuels, U.S. Department of Energy, Washington, DC 20585. January, 2010. Available at: <http://www.eia.doe.gov/cneaf/electricity/epa/epa_sum.html >.

U.S. Environmental Protection Agency (EPA). 2001. *Database of information collected in the Electric Utility Steam Generating Unit Mercury Emissions Information Collection Effort*. OMB Control No. 2060-0396. U.S. Environmental Protection Agency, Office of Air Quality Planning and Standards, Research Triangle Park, NC. April 2001.

U.S. Environmental Protection Agency (EPA). 2005. *Multipollutant Emission Control Technology Options for Coal-fired Power Plants*. EPA. EPA-600/R-05/034. U.S. Environmental Protection Agency, National Risk Management Research Laboratory. Research Triangle Park, NC. March 2005. Available at: <http://www.epa.gov/airmarkt/resource/docs/multireport2005.pdf>.

U.S. Environmental Protection Agency (EPA). 2006. *Environmental Footprints and Costs of Coat-Based Integrated Gasification Combined Cycle and Pulverized Coal Technologies*. EPA. EPA-600/R-06/006. U.S. Environmental Protection Agency, National Risk Management Research Laboratory. Research Triangle Park, NC. July 2006. Available at: <http://www.epa.gov/oar/caaac/coaltech/2007_01_epaigcc.pdf>.

Vattenfall. 2006. *Nordjylland Power Station: The World's Most Efficient Coal-Fired CHP Plant*. Vattenfall, A/S Copenhagen, Denmark. Available at: <http://www.vattenfall.dk/da/file/nordjyllanduk080121_7841588.pdf>.

VGB Powertech and Evonik Energy Services (2008). Supercritical and ultra supercritical technology. In *Power Plant Performance Reporting and Improvement under the Provision of the Indian Energy Conservation Act: Output 1.1: Best practice performance monitoring, analysis of performance procedures, software and analytical tools, measuring instrumentation, guidelines or best practice manuals and newest trends.* (Annexure XIV). Indo German Energy Programme. Retrieved from <http://www.emt-india.net/PowerPlantComponent/Cutput1.1/Output1.1.pdf>.

Weijuan, Yang, et al. 2007. *Nitrous Oxide Formation and Emission in Selective Non-Catalytic Reduction.* Energy Power Engineering China, 228-232.

In: Reducing Greenhouse Gas Emissions   ISBN: 978-1-61470-726-4
Editors: Diane B. McCreevey and Ellen L. Durkin   © 2011 Nova Science Publishers, Inc.

Chapter 2

AVAILABLE AND EMERGING TECHNOLOGIES FOR REDUCING GREENHOUSE GAS EMISSIONS FROM THE PETROLEUM REFINING INDUSTRY[*]

United States Environmental Protection Agency

1.0. INTRODUCTION

This document is one of several white papers that summarize readily available information on control techniques and measures to mitigate greenhouse gas (GHG) emissions from specific industrial sectors. These white papers are solely intended to provide basic information on GHG control technologies and reduction measures in order to assist States and local air pollution control agencies, tribal authorities, and regulated entities in implementing technologies or measures to reduce GHGs under the Clean Air Act, particularly in permitting under the prevention of significant deterioration (PSD) program and the assessment of best available control technology (BACT). These white papers do not set policy, standards or otherwise establish any binding requirements; such requirements are contained in the applicable EPA regulations and approved state implementation plans.

This document provides information on control techniques and measures that are available to mitigate greenhouse gas (GHG) emissions from the petroleum refining industry at this time. Because the primary GHG emitted by the petroleum refining industry are carbon dioxide (CO_2) and methane (CH_4), the control technologies and measures presented here focus on these pollutants. While a large number of available technologies are discussed here, this paper does not necessarily represent all potentially available technologies or measures that that may be considered for any given source for the purposes of reducing its GHG emissions. For example, controls that are applied to other industrial source categories with exhaust streams similar to the petroleum refining industry may be available through "technology transfer" or new technologies may be developed for use in this sector.

---

[*] This is an edited, reformatted and augmented version of the United States Environmental Protection Agency publication, dated October 2010.

The information presented in this document does not represent U.S. EPA endorsement of any particular control strategy. As such, it should not be construed as EPA approval of a particular control technology or measure, or of the emissions reductions that could be achieved by a particular unit or source under review.

2.0. PETROLEUM REFINING

2.1. Overview of Petroleum Refining Industry

Petroleum refineries produce liquefied petroleum gases (LPG), motor gasoline, jet fuels, kerosene, distillate fuel oils, residual fuel oils, lubricants, asphalt (bitumen), and other products through distillation of crude oil or through redistillation, cracking, or reforming of unfinished petroleum derivatives. There are three basic types of refineries: topping refineries, hydroskimming refineries, and upgrading refineries (also referred to as "conversion" or "complex" refineries). Topping refineries have a crude distillation column and produce naphtha and other intermediate products, but not gasoline. There are only a few topping refineries in the U.S., predominately in Alaska. Hydroskimming refineries have mild conversion units such as hydrotreating units and/or reforming units to produce finished gasoline products, but they do not upgrade heavier components of the crude oil that exit near the bottom of the crude distillation column. Some topping/hydroskimming refineries specialize in processing heavy crude oils to produce asphalt. There are eight operating asphalt plants and approximately 20 other hydroskimming refineries operating in the United States as of January 2006 (Energy Information Administration [EIA], 2006). The vast majority (approximately 75 to 80 percent) of the approximately 150 domestic refineries are upgrading/conversion refineries. Upgrading/conversion refineries have cracking or coking operations to convert long-chain, high molecular weight hydrocarbons ("heavy distillates") into smaller hydrocarbons that can be used to produce gasoline product ("light distillates") and other higher value products and petrochemical feedstocks.

Figure 1 provides a simplified flow diagram of a typical refinery. The flow of intermediates between the processes will vary by refinery, and depends on the structure of the refinery, type of crude processes, as well as product mix. The first process unit in nearly all refineries is the crude oil or "atmospheric" distillation unit (CDU). Different conversion processes are available using thermal or catalytic processes, *e.g.*, delayed coking, catalytic cracking, or catalytic reforming, to produce the desired mix of products from the crude oil. The products may be treated to upgrade the product quality (*e.g.*, sulfur removal using a hydrotreater). Side processes that are used to condition inputs or produce hydrogen or by-products include crude conditioning (*e.g.*, desalting), hydrogen production, power and steam production, and asphalt production. Lubricants and other specialized products may be produced at special locations. More detailed descriptions of petroleum refining processes are available in other locations (U.S. EPA, 1995, 1998; U.S. Department of Energy [DOE], 2007).

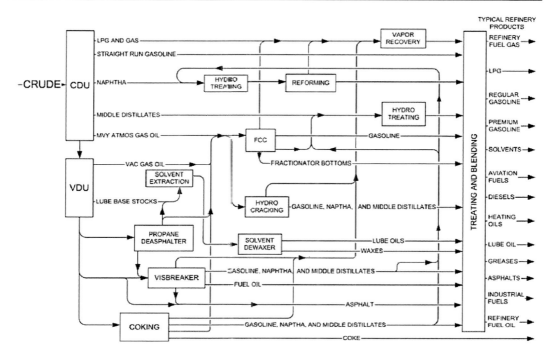

Figure 1. Simplified flowchart of refining processes and product flows. Adapted from Gary and Handwerk (1994).

2.2. Petroleum Refining GHG Emission Sources

The petroleum refining industry is the nation's second-highest industrial consumer of energy (U.S. DOE, 2007). Nearly all of the energy consumed is fossil fuel for combustion; therefore, the petroleum refining industry is a significant source of GHG emissions. In addition to the combustion-related sources (*e.g.*, process heaters and boilers), there are certain processes, such as fluid catalytic cracking units (FCCU), hydrogen production units, and sulfur recovery plants, which have significant process emissions of CO_2. Methane emissions from a typical petroleum refinery arise from process equipment leaks, crude oil storage tanks, asphalt blowing, delayed coking units, and blow down systems. Asphalt blowing and flaring of waste gas also contributes to the overall CO_2 and CH_4 emissions at the refinery. Based on a bottom-up, refinery-specific analysis (adapted from Coburn, 2007, and U.S. EPA, 2008), GHG emissions from petroleum refineries were estimated to be 214-million metric tons of CO_2 equivalents (CO_2e), based on production rates in 2005. Figure 2 provides a breakdown of the nationwide emissions projected for different parts of the petroleum refineries based on this bottom-up analysis.

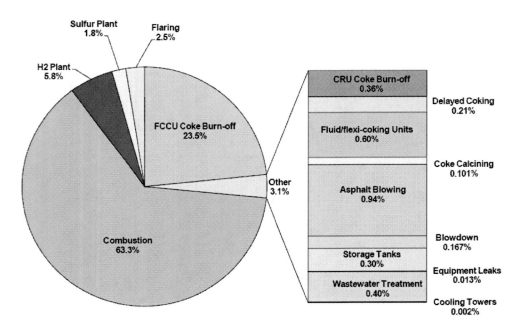

Figure 2. Contribution of different emission sources to the nationwide CO_2 equivalent GHG emissions from petroleum refineries.

Figure 3 presents what GHG are emitted by refineries. CO_2 is the predominant GHG emitted by petroleum refineries, accounting for almost 98 percent of all GHG emissions at petroleum refineries. Methane emissions are 4.7-million metric tons CO_2e and account for 2.25 percent of the petroleum refinery emissions nationwide. Note that the relative magnitude of CO_2 and CH_4 emissions is dependent on the types of process units and other characteristics of the refinery. Facilities that do not have catalytic cracking units and hydrogen plants will tend to have a higher fraction of their total GHG emissions released as CH_4.

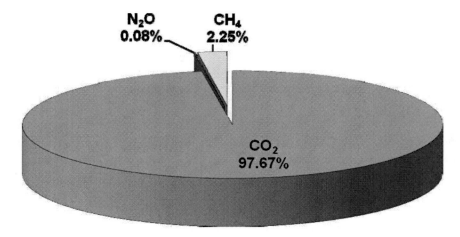

Figure 3. GHG emissions from petroleum refineries.

The petroleum refining industry is one of the largest energy consuming industries in the United States. Thus, a primary option to reduce CO2 emissions is to improve energy efficiency. In 2001, the domestic petroleum refining industry consumed an estimated 3,369 trillion British Thermal Units (TBtu). One report estimated the CO_2 emissions from this energy consumption to be about 222 million tonnes, which accounts for about 11.6 percent of industrial CO_2 emissions in the United States (Worrell and Galitsky, 2005). The EIA provides on-site fuel consumption data as well as electricity and steam purchases (EIA, 2009). These data were used to estimate the CO_2 emissions resulting from this fuel consumption using the emission factors from the Intergovernmental Panel on Climate Change (IPCC) (2006), and converted to appropriate units (Coburn 2007). Figure 4 presents the projected CO_2 emissions from the direct, on-site fuel consumption, as well as the indirect, off-site electricity and steam purchases. From Coburn (2007), the on-site annual CO_2 emissions from fuel combustion were 190 million tonnes in 2005 and the overall CO_2 emissions from energy consumption (including purchased steam and electricity) were 216 million tonnes in 2005, which agrees well with the estimate of Worrell (2005). As seen in Figure 4, catalyst coke consumption dropped 10 percent from 2006 to 2008. Much of the resulting CO_2 emission reductions were offset by increased electricity and steam purchases. As nearly all catalytic cracking units recover the latent heat from the catalyst coke burn-off exhaust to produce steam and/or electricity, the decrease in catalyst coke consumption does not translate into an equivalent net GHG emissions reduction when indirect CO_2 emissions from electricity and steam purchases are considered.

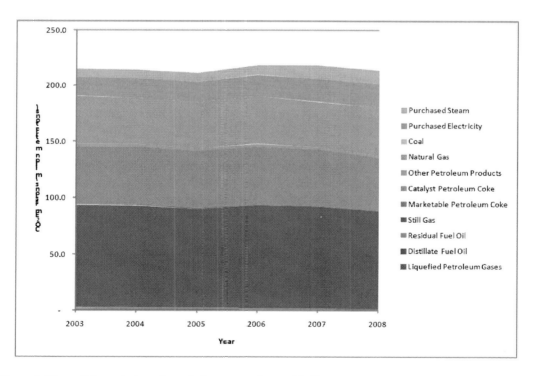

Fgure 4. Direct CO_2 emissions from fuel consumption and indirect CO_2 emissions from electricity and steam purchases at U.S. petroleum refineries from 2003 to 2008.

The remainder of this section provides brief descriptions of the process units and other sources that generate significant GHG emissions at a petroleum refinery.

2.2.1. *Stationary Combustion Sources*

Stationary combustion sources are the largest sources of GHG emissions at a petroleum refinery. Combustion sources primarily emit CO_2, but they also emit small amounts of CH_4 and nitrous oxide (N_2O). Stationary combustion sources at a petroleum refinery include process heaters, boilers, combustion turbines, and similar devices. For this document, flares are considered a distinct emission source separate from other stationary combustion sources. Nearly all refinery process units use process heaters. Typically, the largest process heaters at a petroleum refinery are associated with the crude oil atmospheric and vacuum distillation units and the catalytic reforming unit (if present at the refinery).

In addition to direct process heat, many refinery processes also have steam and electricity requirements. Some refineries purchase steam to meet their process's steam requirements; others use dedicated on-site boilers to meet their steam needs. Similarly, some refineries purchase electricity from the grid to run their pumps and other electrical equipment; other refineries have co-generation facilities to meet their electricity needs and may produce excess electricity to sell to the grid. Refineries that produce their own steam or electricity will have higher on-site fuel usage, all other factors being equal, than refineries that purchase these utilities. A boiler for producing plant steam can be the largest source of GHG emissions at the refinery, particularly at refineries that do not have catalytic cracking units.

The predominant fuel used at petroleum refineries is refinery fuel gas (RFG), which is also known as still gas. RFG is a mixture of light C1 to C4 hydrocarbons, hydrogen, hydrogen sulfide (H_2S), and other gases that exit the top (overhead) of the distillation column and remain uncondensed as they pass through the overhead condenser. RFG produced at different locations within the refinery is typically compressed, treated to remove H_2S (if necessary), and routed to a central location (*i.e.*, mix drum) to supply fuel to the various process heaters at the refinery. This RFG collection and distribution system is referred to as the fuel gas system. A refinery may have several fuel gas systems, depending on the configuration of the refinery, supplying fuel to different process heaters and boilers.

The fuel gas generated at the refinery is typically augmented with natural gas to supply the full energy needs of the refinery. Depending on the types of crude oil processed and the process units in operation, the amount of supplemental natural gas needed can change significantly. Consequently, there may be significant variability in the fuel gas composition between different refineries and even within a refinery as certain units are taken off-line for maintenance.

2.2.2. *Flares*

Flares are commonly used in refineries as safety devices to receive gases during periods of process upsets, equipment malfunctions, and unit start-up and shutdowns. Some flares receive only low flows of "purge" or "sweep" gas to prevent air (oxygen) from entering the flare header and possibly the fuel gas system while maintaining the readiness of the flare in the event of a significant malfunction or process upset. Some flares may receive excess process gas on a frequent or routine basis. Some flares may be used solely as control devices for regulatory purposes. Combustion of gas in a flare results in emissions of predominately CO_2, along with small amounts of CH_4 and N_2O.

2.2.3. Catalytic Cracking Units

In the catalytic cracking process, heat and pressure are used with a catalyst to break large hydrocarbons into smaller molecules. The FCCU is the most common type of catalytic cracking unit currently in use. The FCCU feed is pre-heated to between 500 and 800 degrees Fahrenheit (°F) and contacted with fine catalyst particles from the regenerator section, which are at about 1,300 °F in the feed line ("riser"). The feed vapor, which is heavy distillate oil from the crude or vacuum distillation column, reacts when contacted with the hot catalyst to break (or crack) the large hydrocarbon compounds into a variety of lighter hydrocarbons. During this cracking process, coke is deposited on the catalyst particles, which deactivates the catalyst. The catalyst separates from the reacted ("cracked") vapors in the reactor; the vapors continue to a fractionation tower and the catalyst is recycled to the regenerator portion of the FCCU to burn-off the coke deposits and prepare the catalyst for reuse in the FCCU riser/reactor (U.S. EPA, 1998).

The FCCU catalyst regenerator generates GHG through the combustion of coke (essentially solid carbon with small amounts of hydrogen and various impurities) that was deposited on the catalyst particles during the cracking process. CO_2 is the primary GHG emitted; small quantities of CH_4 and N_2O are also emitted during "coke burn-off." An FCCU catalyst regenerator can be designed for complete or partial combustion. A complete-combustion FCCU operates with sufficient air to convert most of the carbon to CO_2 rather than carbon monoxide (CO). A partial-combustion FCCU generates CO as well as CO_2, so most partial-combustion FCCU are typically followed by a CO boiler to convert the CO to CO_2. Most refineries that operate an FCCU recover useful heat generated from the combustion of catalyst coke during catalyst regeneration; the heat recovered from catalyst coke combustion offsets some of the refinery's ancillary energy needs. The FCCU catalyst regeneration or coke burn-off vent is often the largest single source of GHG emissions at a refinery.

Thermal catalytic cracking units (TCCU) are similar to FCCU, except that the catalyst particles are much bigger and the system uses a moving bed reactor rather than a fluidized system. The generation of GHG, however, is the same. Specifically, GHG are generated in the regenerator section of the TCCU when coke deposited on the catalyst particles is burned-off in order to restore catalyst activity.

2.2.4. Coking Units

Coking is another cracking process, usually used at a refinery to generate transportation fuels, such as gasoline and diesel, from lower-value fuel oils. A desired by-product of the coking reaction is petroleum coke, which can be used as a fuel for power plants as well as a raw material for carbon and graphite products. Coking units are often installed at existing refineries to increase the refinery's ability to process heavier crude oils. There are three basic types of coking units: delayed coking units, (traditional) fluid coking units, and flexicoking units. Delayed coking units are the most common, and all new coking units are expected to be delayed cokers.

Delayed Coking Units

Delayed coking is a semibatch process using two coke drums and a single fractionator tower (distillation column) and coking furnace. A feed stream of heavy residues is introduced to the fractionating tower. The bottoms from the fractionator are heated to about 900 to 1,000

°F in the coking furnace, and then fed to an insulated coke drum where thermal cracking produces lighter (cracked) reaction products and coke. The reaction products produced in the coke drum are fed back to the fractionator for product separation. After the coke drum becomes filled with coke, the feed is alternated to the parallel (empty) coke drum, and the filled coke drum is purged and cooled, first by steam injection, and then by water addition. A coke drum blowdown system recovers hydrocarbon and steam vapors generated during the quenching and steaming process. Once cooled, the coke drum is vented to the atmosphere, opened, and then high pressure water jets are used to cut the coke from the drum. After the coke cutting cycle, the drum is closed and preheated to prepare the vessel for going back on-line (i.e., receiving heated feed). A typical coking cycle will last for 16 to 24 hours on-line and 16 to 24 hours cooling and decoking. The primary GHG released from a delayed coking unit is CH_4, which is emitted both from the blowdown system (if not controlled) and from the atmospheric venting and opening of the coke drum.

Fluid Coking Units

The fluid coking process is continuous and occurs in a reactor rather than a coke drum like the delayed coking process. Fluid coking units produce a higher grade of petroleum coke than delayed coking units; however, unlike delayed coking units that use large process preheaters, fluid coking units burn 15 to 25 percent of the coke produced to provide the heat needed for the coking reactions (U.S. DOE, 2007). The coke is burned with limited air, so large quantities of CO are produced (similar to a partial combustion FCCU), which are subsequently burned in a CO boiler. Like the FCCU, the combustion of the petroleum coke and subsequent combustion of CO generates large quantities of CO_2 along with small amounts of CH_4 and N_2O. For the few refineries with fluid coking units, the fluid coking units are significant contributors to the refinery's GHG emissions. Fluid coking units are not significant contributors to the nationwide emissions totals because there are only three fluid coking units in the United States; however, fluid coking units have emissions comparable to (and slightly greater than) catalytic cracking units of the same throughput capacity.

Flexicoking Units

The flexicoking process is very similar to the fluid coking unit except that a coke gasifier is added that burns nearly all of the produced coke at $1700 - 1800$ °F with steam to produce low heating value synthesis gas (syngas). The produced syngas, along with entrained fines, is routed through the heater vessel for fluidization of the hot coke bed and for heat transfer to the solids. The syngas is then treated to remove entrained particles and reduced sulfur compounds and the syngas can then be used in specially designed boilers or other combustion sources that can accommodate the low heat content of the syngas. Most of the CO_2 emissions produced in the flexicoking unit will not be released at the unit, but rather it will be part of the syngas. Some of the CO_2 produced in the flexicoking unit is expected to be removed as part of the sulfur removal process and subsequently released in the sulfur recovery plant; the CO_2 that remains in the scrubbed syngas will be released from the stationary combustion unit that uses the syngas as fuel (usually a boiler specifically designed to use the low heating value content syngas). Therefore, while the flexicoking unit is not expected to have significant GHG emissions directly from the unit, the flexicoking unit will impact the energy balance and GHG emissions from other sources at the refinery.

2.2.5. Catalytic Reforming Units

In the catalytic reforming unit (CRU), low-octane hydrocarbon distillates, generally gasoline and naphtha are reacted with a catalyst to produce aromatic compounds such as benzene. An important by-product of the reforming reaction is hydrogen. The feed to the CRU must be treated to remove sulfur, nitrogen, and metallic contaminants, typically using a catalytic hydrotreater (which will consume some hydrogen, but not as much as produced in the CRU). The CRU usually has a series of three or four reactors. The reforming reactor is endothermic, so the feed must be heated prior to each reactor vessel. Coke deposits slowly on the catalyst particles during the processing reaction, and this "catalyst coke" must be burned-off to reactivate the catalyst, generating CO_2, along with small amounts of CH_4 and N_2O.

There are three types of CRU based on how the regeneration of the catalyst is performed: continuous CRU, cyclic CRU, and semi-regenerative CRU. In a continuous CRU (or platformers), small quantities of the catalyst are continuously removed from a moving bed reactor system, purged, and transported to a continuously operated regeneration system. The regenerated catalyst is then recycled to the moving bed reactor. Continuous reformers generally operate at lower pressures than other reforming units, resulting in higher coke deposition rates. Cyclic CRU has an extra reactor vessel, so that one reactor vessel can be isolated from the unit for regeneration. After the first vessel is regenerated, it is brought back on-line and the second reactor vessel is then isolated and regenerated and so on until all vessels have been regenerated. Thus, in cyclic units, the CRU continues to operate and individual reactor vessels are regenerated in a cyclical process many times during a single year. In a semi-regenerative CRU, the entire reforming unit is taken off-line to regenerate the catalyst in the reactor vessels. Catalyst regeneration in a semi-regenerative CRU typically occurs once every 12 to 24 months (18 months is typical) and lasts approximately 1 to 2 weeks (U.S. EPA, 1998).

In addition to the CO_2 generated during coke burn-off, there may be some CH_4 emissions during the depressurization and purging of the reactor vessels of recycled catalyst prior to regeneration. While the CH_4 emissions from the depressurization and purging processes are expected to be negligible in most cases, natural gas (*i.e.*, CH_4) is occasionally used as the purge gas, in which case the CH_4 emissions would not be negligible.

2.2.6. Sulfur Recovery Vents

Hydrogen sulfide (H_2S) is removed from the refinery fuel gas system through the use of amine scrubbers. While the selectivity of H_2S removal is dependent on the type of amine solution used, these scrubbers also tend to extract CO_2 from the fuel gas. The concentrated sour gas is then processed in a sulfur recovery plant to convert the H_2S into elemental sulfur or sulfuric acid. CO_2 in the sour gas will pass through the sulfur recovery plant and be released in the final sulfur plant vent. Additionally, small amounts of hydrocarbons may also be present in the sour gas stream. These hydrocarbons will eventually be converted to CO_2 in the sulfur recovery plant or via tail gas incineration. The most common type of sulfur recovery plant is the Claus unit, which produces elemental sulfur. The first step in a Claus unit is a burner to convert one-third of the sour gas into sulfur dioxide (SO_2) prior to the Claus catalytic reactors. GHG emissions from the fuel fired to the Claus burner are expected to be accounted for as a combustion source. After that, the sulfur dioxide and unburned H_2S are reacted in the presence of a bauxite catalyst to produce elemental sulfur. Based on process-specific data collected in the development of emission standards for petroleum refineries, there are 195 sulfur recovery trains in the petroleum refining industry (U.S. EPA, 1998).

2.2.7. Hydrogen Plants

The most common method of producing hydrogen at a refinery is the steam methane reforming (SMR) process. Methane, other light hydrocarbons, and steam are reacted via a nickel catalyst to produce hydrogen and CO. Excess CH_4 is added and combusted to provide the heat needed by this endothermic reaction. The CO generated by the initial reaction further reacts with the steam to generate hydrogen and CO_2 (U.S. DOE, 2007). According to EIA's Refinery Capacity Report 2006 (EIA, 2006), 54 of the 150 petroleum refineries have hydrogen production capacity. CO_2 produced as a byproduct of SMR hydrogen production accounts for approximately 6 percent of GHG emissions from petroleum refineries nationwide, but can account for 25 percent of the GHG emissions from an individual refinery. Many of the hydrogen plants located at a petroleum refinery are operated by third-parties. It is unclear if the hydrogen production units reported by EIA include all hydrogen plants co-located at a refinery or only those that are directly owned and operated by the refinery.

2.2.8. Asphalt Blowing Stills

Asphalt or bitumen blowing is used for polymerizing and stabilizing asphalt to improve its weathering characteristics in the production of asphalt roofing products and certain road asphalts. Asphalt blowing involves the oxidation of asphalt flux by bubbling air through liquid asphalt flux at 260 °Celsius (C) (500 °F) for 1 to 10 hours depending on the desired characteristics of the product. The vessel used for asphalt blowing is referred to as a "blowing still." The emissions from a blowing still are primarily organic particulate with a fairly high concentration of gaseous hydrocarbon and polycyclic organic matter as well as reduced sulfur compounds. The blowing still gas also contains significant quantities of CH_4 and CO_2. The blowing still gas is commonly controlled with a wet scrubber to remove sour gas, entrained oil, particulates, and condensable organics and/or a thermal oxidizer to combust the hydrocarbons and sour gas to CO_2 and SO_2.

2.2.9. Storage Tanks

Storage tanks will generally have negligible GHG emissions except when unstabilized crude oil is stored or a methane blanket is used in the storage tank. Unstabilized crude oil is crude oil that has not been stored at atmospheric conditions for prolonged periods of time (several days to a week) prior to being received at the refinery. Most crude oil deposits also include natural gas (i.e., CH_4); some of the CH_4 is dissolved in the crude oil at the pressure of the crude oil deposit. When crude oil is extracted, it is often stored temporarily at atmospheric conditions to either discharge or recover the dissolved gases. If the crude oil is transported under pressure (e.g., via a pipeline) either immediately or shortly after extraction, the dissolved gases will remain in the crude oil until it reaches the refinery. The dissolved gases will be subsequently released from this "unstabilized" crude oil when the crude oil is stored at atmospheric conditions at a storage tank at the refinery.

2.2.10. Coke Calcining Units

Coke calcining units are a significant source of CO_2 emissions; however, only a few petroleum refineries have on-site coke calcining units. Coke calciners are used to burn-off sulfur, volatiles, and other impurities in the coke to produce a premium grade coke that can be used to make electrodes, anode vessels, and other products. A small fraction of the coke is consumed/pyrolyzed in the process under oxygen starved conditions; the process gas generated is then combusted in an afterburner by mixing the process gas with air in the presence of a flame. Most of the CO_2 generated from the process/afterburner system is attributable to the volatile content of the coke fed to the calciner.

2.2.11. Other Ancillary Sources

Refineries may also contain other ancillary sources of GHG emissions. Most refineries have wastewater treatment systems and some refineries have landfills. While the aerobic biodegradation of wastes is generally considered to be biogenic, anaerobic degradation of waste producing CH_4 emission is not. The high organic loads and stagnant conditions in an oil-water separator are conducive to anaerobic degradation, and the oil water separator may be a fairly significant ancillary source of CH_4 emissions. Landfills are also conducive to anaerobic degradation. Depending on the organic content of the waste material managed in a landfill, the landfill may also be a fairly significant ancillary source of CH_4 emissions.

The refinery's fuel gas system will generally contain significant concentrations of CH4; certain process units may either generate methane or use methane and other light ends as part of the process operations (*e.g.*, SMR hydrogen production). Leaking equipment components (*e.g.*, valves, pumps, and flanges) may, therefore, be a source of CH_4 emissions. Leak detection and repair (LDAR) programs are commonly used to identify and reduce emissions from equipment components; however most LDAR programs exclude the fuel gas system. Similar to equipment leaks, some heat exchangers may develop leaks whereby gases being cooled can leak into the cooling water. Although these leaks are not direct releases to the atmosphere, light hydrocarbons that leak into the cooling water will generally be released to the atmosphere in cooling towers (for recirculated cooling water systems) or ponds/receiving waters (in once through systems). As several heat exchangers at a refinery cool gases that contain appreciable quantities of CH_4 (e.g., a distillation column's overhead condenser), cooling towers may also be a source of CH4 emissions. Nonetheless, CH4 emissions from equipment leaks, either directly to the atmosphere from leaking equipment components or indirectly from cooling towers from leaking heat exchangers, are generally expected to have a minimal contribution to a typical refinery's total GHG emissions.

3.0. SUMMARY OF GHG REDUCTION MEASURES

Table 1 summarized the GHG reduction measures described in this document. Additional detail regarding these GHG reduction measures are provided in Section 4, Energy Programs and Management Systems, and Section 5, GHG Reduction Measures by Source, of this document.

Table 1. Summary of GHG Reduction Measures for the Petroleum Refining Industry

GHG Control Measure	Description	Efficiency Improvement/ GHG emission reduction	Retrofit Capital Costs ($/unit of CO2e)	Payback time (years)	Demonstrated in Practice?	Other Factors
Energy Efficiency Programs and Systems						
	Benchmark GHG performance and implement energy management systems to improve energy efficiency, such as:					
Energy Efficiency Initiatives and Improvements	■ improve process monitoring and control systems ■ use high efficiency motors ■ use variable speed drives ■ optimize compressed air systems ■ implement lighting system efficiency improvements	4-17% of electricity consumption		1-2 years	Yes	
Stationary Combustion Sources						
Steam Generating Boilers (see also ICI Boiler GHG BACT Document)						

GHG Control Measure	Description	Efficiency Improvement/ GHG emission reduction	Retrofit Capital Costs ($/unit of	Payback time (years)	Demonstrated in Practice?	Other Factors
Systems Approach to Steam Generation	Analyze steam needs and energy recovery options, including: ■ minimize steam generation at excess pressure or volume ■ use turbo or steam expanders when excesses are unavoidable ■ schedule boilers based on efficiency				Yes	
Boiler Feed Water Preparation	Replace a hot lime water softener with a reverse osmosis membrane treatment system to remove hardness and reduce alkalinity of boiler feed.	70-90% reduction in blowdown steam loss; up to 10% reduction in GHG emissions		2-5 years	Yes	
Improved Process Control	Oxygen monitors and intake air flow monitors can be used to optimize the fuel/air mixture and limit excess air.	1-3% of boiler emissions		6 - 18 months	Yes	Low excess air levels may increase CO emissions.
Improved Insulation	Insulation (or improved insulation) of boilers and distribution pipes.	3-13% of boiler emissions		6 - 18 months	Yes	

Table 1. (Continued)

GHG Control Measure	Description	Efficiency Improvement/ GHG emission reduction	Retrofit Capital Costs ($/unit of CO_2e)	Payback time (years)	Demonstrated in Practice?	Other Factors
Improved Maintenance	All boilers should be maintained according to a maintenance program. In particular, the burners and condensate return system should be properly adjusted and worn components replaced. Additionally, fouling on the fireside of the boiler and scaling on the waterside should be controlled.	1-10% of boiler emissions			Yes	
Recover Heat from Process Flue Gas	Flue gases throughout the refinery may have sufficient heat content to make it economical to recover the heat. Typically, this is accomplished using an economizer to preheat the boiler feed water.	2-4% of boiler emissions		2 years	Yes	
Recover Steam from Blowdown	Install a steam recover system to recover blowdown steam for low pressure steam needs (e.g., space heating and feed water preheating).	1 –3%		1 - 3 years	Yes	

GHG Control Measure	Description	Efficiency Improvement/ GHG emission reduction	Retrofit Capital Costs ($/unit of CO_2e)	Payback time (years)	Demonstrated in Practice?	Other Factors
Reduce Standby Losses	Reduce or eliminate steam production at standby by modifying the burner, combustion air supply, and boiler feedwater supply, and using automatic control systems to reduce the time needed to reach full boiler capacity.	Up to 85% reduction in standby losses (but likely a small fraction of facility total boiler emissions)		1.5 years	Yes	
Improve and Maintain Steam Traps	Implement a maintenance plan that includes regular inspection and maintenance of steam traps to prevent steam lost through malfunctioning steam traps.	1-10% of boiler emissions			Yes	
Install Steam Condensate Return Lines	Reuse of the steam condensate reduces the amount of feed water needed and reduces the amount of energy needed to produce steam since the condensate is preheated.	1- 10% of steam energy use		1-2 years	Yes	
Process Heaters						
Combustion Air Controls- Limitations on Excess air	Oxygen monitors and intake air flow monitors can be used to optimize the fuel/air mixture and limit excess air.	1-3%		6-18 months	Yes	

Table 1. (Continued)

GHG Control Measure	Description	Efficiency Improvement/ GHG emission reduction	Retrofit Capital Costs ($/unit of CO_2e)	Payback time (years)	Demonstrated in Practice?	Other Factors
Heat Recovery: Air Preheater	Air preheater package consists of a compact air-to-air heat exchanger installed at grade level through which the hot stack gases from the convective section exchange heat with the incoming combustion air. If the original heater is natural draft, a retrofit requires conversion to mechanical draft.	10-15% over units with no preheat.			Yes	**May increase NOx emissions**
Combined Heat and Power						
Combined Heat and Power	Use internally generated fuels or natural gas for power (electricity) production using a gas turbine and generate steam from waste heat of combustion exhaust to achieve greater energy efficiencies			5 years	Yes	
Carbon Capture						
Oxy-combustion	Use pure oxygen in large combustion sources to reduce flue gas volumes and increase CO_2 concentrations to improve capture efficiency and costs				No	

GHG Control Measure	Description	Efficiency Improvement/ GHG emission reduction	Retrofit Capital Costs ($/unit of CO_2e)	Payback time (years)	Demonstrated in Practice?	Other Factors
Post-combustion Solvent Capture	Use solvent scrubbing, typically using monoethanolamine (MEA) as the solvent, for separation of CO_2 in post-combustion exhaust streams				Yes	
Post-combustion membranes	Use membrane technology to separate or adsorb CO_2 in an exhaust stream		$55-63		No	
Fuel Gas System and Flares						
Fuel Gas System						
Compressor Selection	Use dry seal rather than wet seal compressors; use rod packing for reciprocating compressors				Yes	
Leak Detection and Repair	Use organic vapor analyzer or optical sensing technologies to identify leaks in natural gas lines, fuel gas lines, and other lines with high methane concentrations and repair the leaks as soon as possible.	80-90% of leak emissions; <0.1% refinery-wide			Yes	
Sulfur Scrubbing System	Evaluate different sulfur scrubbing technologies or solvents for energy efficiency				Yes	

Table 1. (Continued)

GHG Control Measure	Description	Efficiency Improvement/ GHG emission reduction	Retrofit Capital Costs ($/unit of CO_2e)	Payback time (years)	Demonstrated in Practice?	Other Factors
Flares						
Flare Gas Recovery	Install flare gas recovery compressor system to recover flare gas to the fuel gas system			1 yr	Yes	
Proper Flare Operation	Maintain combustion efficiency of flare by controlling heating content of flare gas and steam- or air-assist rates				Yes	
Refrigerated Condensers	Use refrigerated condensers to increase product recovery and reduce excess fuel gas production				Yes	
Cracking Units						
Fluid Catalytic Cracking Units (see also: Stationary Combustion Sources; Fuel Gas System and Flares)						
Power/Waste Heat Recovery	Install or upgrade power recovery or waste heat boilers to recover latent heat from the FCCU regenerator exhaust				Yes	
High-Efficiency Regenerators	Use specially designed FCCU regenerators for high efficiency, complete combustion of catalyst coke deposits				Yes	

GHG Control Measure	Description	Efficiency Improvement/ GHG emission reduction	Retrofit Capital Costs ($/unit of CO_2e)	Payback time (years)	Demonstrated in Practice?	Other Factors
Hydrocracking Units (see also: Stationary Combustion Sources; Fuel Gas System and Flares; Hydrogen Production Units)						
Power/Waste Heat Recovery	Install or upgrade power recovery to recover power from power can be recovered from the pressure difference between the reactor and fractionation stages			2.5 years	Yes	
Hydrogen Recovery	Use hydrogen recovery compressor and back-up compressor to ensure recovery of hydrogen in process off-gas				Yes	
Coking Units						
Fluid Coking Units (see also: Stationary Combustion Sources; Fuel Gas System and Flares)						
Power/Waste Heat Recovery	Install or upgrade power recovery or waste heat boilers to recover latent heat from the fluid coking unit exhaust				Yes	
Flexicoking Units (see: Stationary Combustion Sources; Fuel Gas System and Flares)						
Delayed Coking Units (see also: Stationary Combustion Sources; Fuel Gas System and Flares)						
Steam Blowdown System	Use low back-pressure blowdown system and recycle hot blowdown system water for steam generation				Yes	
Steam Vent	Lower pressure and temperature of coke drum to 2 to 5 psig and 230°F to minimize direct venting emissions	50 to 80% reduction in direct steam vent CH_4 emissions			Yes	

Table 1. (Continued)

GHG Control Measure	Description	Efficiency Improvement/ GHG emission reduction	Retrofit Capital Costs ($/unit of CO2e)	Payback time (years)	Demonstrated in Practice?	Other Factors
Catalytic Reforming Units (see also: Stationary Combustion Sources; Fuel Gas System and Flares; Hydrogen Production Units)						
Sulfur Recovery Units						
Sulfur Recovery System Selection	Evaluate energy and CO2 intensity in selection of sulfur recovery unit and tail gas treatment system and a variety of different tail gas treatment units including Claus, SuperClaus® and EuroClaus®, SCOT, Beavon/amine, Beavon/Stretford, Cansolv®, LoCat®, and Wellman-Lord				Yes	
Hydrogen Production Units						
Hydrogen Production Optimization	Implement a comprehensive assessment of hydrogen needs and consider using additional catalytic reforming units to produce H2				Yes	
Combustion Air and Feed/Steam Preheat	Use heat recovery systems to preheat the feed/steam and combustion air temperature	5% of total energy consumption for H2 production			Yes	

GHG Control Measure	Description	Efficiency Improvement/ GHG emission reduction	Retrofit Capital Costs ($/unit of CO_2e)	Payback time (years)	Demonstrated in Practice?	Other Factors
Cogeneration	Use cogeneration of hydrogen and electricity: hot exhaust from a gas turbine is transferred to the reformer furnace; the reformer convection section is also used as a heat recovery steam generator (HRSG) in a cogeneration design; steam raised in the convection section can be put through either a topping or condensing turbine for additional power generation				Yes	
Hydrogen Purification	Evaluate hydrogen purification processes (i.e., pressure-swing adsorption, membrane separation, and cryogenic separation) for overall energy intensity and potential CO_2 recovery.				Yes	
Hydrotreating Units (see also: Hydrogen Production Units; Sulfur Recovery Units)						
Hydrotreater Design	Use energy efficient hydrotreater designs and new catalyst to increase sulfur removal.				Yes	

Table 1. (Continued)

GHG Control Measure	Description	Efficiency Improvement/ GHG emission reduction	Retrofit Capital Costs ($/unit of CO_2e)	Payback time (years)	Demonstrated in Practice?	Other Factors
Crude Desalting and Distillation Units						
Desalter Design	Alternative designs for the desalter, such as multi-stage units and combinations of AC and DC fields, may increase efficiency and reduce energy consumption.				Yes	
Progressive Distillation Design	Progressive distillation process uses as series of distillation towers working at different temperatures to avoid superheating lighter fractions of the crude oil.	30% reduction in crude heater emissions; 5% or more refinery-wide			Yes	
Storage Tanks						
Vapor Recovery or Control for Unstabilized Crude Oil Tanks	Consider use of a vapor recovery or control system for crude oil storage tanks that receive crude oil that has been stored under pressure ("unstabilized" crude oil)	90-95% reduction in CH_4 from these tanks			Yes	
Heated Storage Tank Insulation	Insulate heated storage tanks				Yes	

4.0. ENERGY PROGRAMS AND MANAGEMENT SYSTEMS

Industrial energy efficiency can be greatly enhanced by effective management of the energy use of operations and processes. U.S. EPA's ENERGY STAR Program works with hundreds of manufacturers and has seen that companies and sites with stronger energy management programs gain greater improvements in energy efficiency than those that lack procedures and management practices focused on continuous improvement of energy performance.

Energy Management Systems (EnMS) provide a framework for managing energy and promote continuous improvement. The EnMS provides the structure for an energy program and its energy team. EnMS establish assessment, planning, and evaluation procedures which are critical for actually realizing and sustaining the potential energy efficiency gains of new technologies or operational changes.

Energy management systems promote continuous improvement of energy efficiency through:

- Organizational practices and policies,
- Team development
- Planning and evaluation,
- Tracking and measurement,
- Communication and employee engagement, and
- Evaluation and corrective measures.

For nearly 10 years, the U.S. EPA s ENERGY STAR Program has promoted an energy management system approach. This approach, outlined in Figure 5, outlines the basic steps followed by most energy management systems approaches.

In recent years, interest in energy management system approaches has been growing. There are many reasons for the greater interest. These include recognition that a lack of management commitment is an important barrier to increasing energy efficiency. Lack of an effective energy team and an effective program result in poor implementation of new technologies and poor implementation of energy assessment recommendations. Poor energy management practices that fail to monitor performance do not ensure that new technologies and operating procedures will achieve their potential to improve efficiency.

Approaches to implementing energy management systems vary. EPA's ENERGY STAR Guidelines for Energy Management are available for public use on the web and provide extensive guidance (see: www.energystar.gov/guidelines). Alternatively, energy management standards are available for purchase from ANSI, ANSI MSE 2001:200 and in the future from ISO, ISO 50001.

While energy management systems can help organizations achieve greater savings through a focus on continuous improvement, they do not guarantee energy savings or CO_2 reductions alone. Combined with effective plant energy benchmarking and appropriate plant improvements, energy management systems can help achieve greater savings.

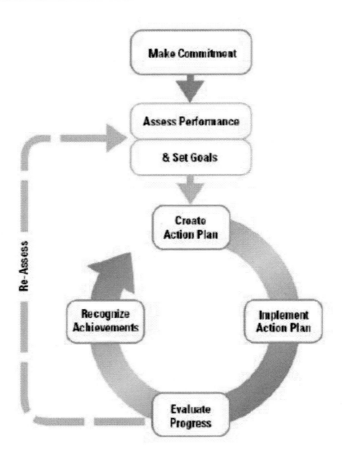

(www.energystar.gov/guidelines)

Figure 5. ENERGY STAR Guidelines for Energy Management.

There are a variety of factors to consider when contemplating requiring certification to an Energy Management Standard established by a standards body such as ANSI or ISO. First, energy management system standards are designed to be flexible. A user of the standard is able to define the scope and boundaries of the energy management system so that single production lines, single processes, a plant or a corporation could be certified. Beyond scope, achieving certification for the first time is not based on efficiency or savings (although re-certifications at a later time could be). Finally, cost is an important factor in the standardized approach. Internal personnel time commitments, external auditor and registry costs are expensive.

From a historical perspective, few companies have pursued certification according to the ANSI energy management standards to date. One reason for this is that the elements of an energy management system can be applied without having to achieve certification which adds additional costs. The ENERGY STAR Guidelines and associated resources are widely used and adopted partly because they are available in the public domain and do not involve certification.

Overall, a systems approach to energy management is an effective strategy for encouraging energy efficiency in a facility or corporation. The focus of energy management efforts are shifted from a "projects" to a "program" approach. There are multiple pathways available with a wide range of associated costs (ENERGY STAR energy management resources are public while the standardized approaches are costly). The effectiveness of an energy management system is linked directly to the system's scope, goals and measurement and tracking. Benchmarks are the most effective measure for demonstrating the system's achievements.

4.1. Sector-Specific Plant Performance Benchmarks

Benchmarking is the process of comparing the performance of one site against itself over time or against the range of performance of the industry. Benchmarking is typically done at a whole facility or site level to capture the synergies of different technologies, operating practices, and operating conditions and typically results in a calculation of the emissions intensity of a site, which are the emissions per unit of product.

For a refinery, emissions intensity is influenced by a number of factors, including energy efficiency, fuel use, feed composition, and products. While refineries all refine crude oil to make a range of common products (gasoline, diesel, fuel oils, liquefied petroleum gases), they often vary in size and the number of processing units that are operating. For example, refineries with more simple configurations may not be able to process certain fractions into more energy-intensive products. Likewise, refineries that process heavy sour crudes may require more energy intensive processing. Benchmarking approaches have been used in the refining industry for many years to improve efficiency and productivity. The European Union evaluated and concluded that the Solomon's complexity weighted barrel approach should be used to benchmark refineries as part of their methodology for allocating emission allowances in the European Union Emissions Trading System (Ecofys, 2009).

4.2. Industry Energy Efficiency Initiatives

The U.S. EPA's ENERGY STAR Program (www.energystar.gov/industry) and U.S. DOE's Industrial Technology Program (www.energy have led industry specific energy efficiency initiatives over the years. These programs have helped to create guidebooks of energy efficient technologies, profiles of industry energy use, and studies of future technologies. Some states have also led sector specific energy efficiency initiatives. Resources from these programs can help to identify technologies that may help reduce CO_2 emissions.

EPA's ENERGY STAR Program has conducted an energy efficiency improvement assessment for petroleum refineries (Worrell and Galitsky, 2005). Many of the GHG reduction measures provided in the following sections are a result of this industry-specific assessment.

4.3. Energy Efficiency Improvements in Facility Operations

4.3.1. Monitoring and Process Control Systems

Most refineries already employ some energy management systems. At existing facilities, only a limited number of processes or energy streams may be monitored and managed. Opportunities should be evaluated for expanding the coverage of monitoring systems throughout the plant. New facilities should include a comprehensive energy management program (Worrell and Galitsky, 2005).

Process control systems are available for essentially all industrial processes. These control systems are typically designed to primarily improve productivity, product quality, and efficiency of a process. However, each of these improvements will lead to increased energy efficiency as well. Process control systems also reduce downtime, maintenance costs, and processing time, and increase resource efficiency and emission control (Worrell and Galitsky, 2005).

Although specific energy savings and payback periods are highly facility-specific, the application of monitoring systems to specific industrial applications have demonstrated energy savings of 4-17 percent, and process control systems can reduce energy consumption by 2-18 percent over facilities without such systems. In general, cost and energy savings of about 5 percent can be expected through the implementation of monitoring and process control systems (Worrell and Galitsky, 2005).

Valero and AspenTech have developed a system to model and control plant-wide energy usage for refinery operations. The system was installed at a domestic refinery and is expected to reduce overall energy usage by 2-8 percent (Worrell and Galitsky, 2005).

Process control systems for the CDU have been shown to reduce energy costs by $0.05-0.12/barrel (bbl) of feed, with paybacks of less than 6 months. Another CDU control system reduced energy consumption and flaring and increased throughput, resulting in a payback of about 1 year. In Portugal, a refinery installed advanced CDU controls and realized a 3-6 percent increase in throughput. The payback period for this control system was 3 months (Worrell and Galitsky, 2005).

Process control systems for FCCU are supplied by several companies. Cost savings range from $0.02-0 40.bbl of feed with paybacks ranging from 6-18 months. At one refinery, an existing FCCU control system was updated at a 65,000 bpd unit and a cost savings of $0.05/bbl of feed was realized. A refinery in Italy installed a control system on a FCCU and reduced cost by $0.10/bbl of feed with a payback of less than 1 year. (Worrell Galitsky, 2005)

In South Africa, a refinery installed a multivariable predictive control system on a hydrotreater. Hydrogen consumption was reduced by 12 percent and the fuel consumption of the heater was reduced by 18 percent. Improved yield of gasoline and diesel were also realized. The payback period was 2 months (Worrell and Galitsky, 2005).

4.3.2. High Efficiency Motors

Electric motors are used throughout the refinery for such applications as pumps, air compressors, fans, and other applications. Pumps, compressors and fans account for 70 to 80 percent of the total electricity usage at the refinery (Worrell and Galitsky, 2005). As such, a systems approach to energy efficiency should be considered for all motor systems (motors, drives, pumps, fans, compressors, controls). An evaluation of energy supply and energy demand could be performed to optimize overall performance. A systems approach includes a motor management plan that considers at least the following factors (Worrell and Galitsky, 2008):

- Strategic motor selection
- Maintenance
- Proper size
- Adjustable speed drives
- Power factor correction
- Minimize voltage unbalances

Pumps are the single largest electricity user at a refinery, accounting for about half of the total energy usage. One study estimated that 20 percent of the energy consumed by pump motors could be saved through equipment or control system changes. Implementation of maintenance programs for pump motors can reduce electricity use by 2-7 percent, with payback periods less than 1 year (Worrell and Galitsky, 2005).

Motor management plans and other efficiency improvements can be implemented at existing facilities and should be considered in the design of new construction. At existing facilities, replacing older motors with high efficiency motors are typically cost-effective when a motor needs replacement, but may not be economical when the old motor is still operational. Payback periods from energy savings are typically less than 1 year (Worrell and Galitsky, 2005).

4.3.3. Variable Speed Drives

Energy use on centrifugal systems such as pumps, fans, and compressors is approximately proportional to the cube of the flow rate. Therefore, small reductions in the flow may result in large energy savings. The use of variable speed drives can better match speed to load requirements of the motors. The installation of variable speed drives at new facilities can result in payback periods of just over 1 year (Worrell and Galitsky, 2005).

4.3.4. Optimization of Compressed Air Systems

Compressed air systems provide compressed air that is used throughout the refinery. Although the total energy used by compressed air systems is small compared to the facility as a whole, there are opportunities for efficiency improvements that will save energy. Efficiency improvements are primarily obtained by implementing a comprehensive maintenance plan for the compressed air systems. Worrell and Galitsky (2005, 2008) listed the following elements of a proper maintenance plan:

- Keep the surfaces of the compressor and intercooling surfaces clean
- Keep motors properly lubricated and cleaned
- Inspect drain traps
- Maintain the coolers
- Check belts for wear
- Replace air lubricant separators as recommended
- Check water cooling systems

In addition to the maintenance plan, reducing leaks in the system can reduce energy consumption by 20 percent. Reducing the air inlet temperature will reduce energy usage, and routing the air intake to outside the building can have a payback in 2-5 years. Control systems can reduce energy consumption by as much as 12 percent. Properly sized pipes can reduce energy consumption by 3 percent. Since as much as 93 percent of the electrical energy used by air compressor systems is lost as heat, recovery of this heat can be used for space heating, water heating, and similar applications (Worrell and Galitsky, 2005, 2008).

Air compressor system maintenance plans and other efficiency improvements can be implemented at existing facilities and should be considered in the design of new construction.

4.3.5. Lighting System Efficiency Improvements

Similar to air compressor systems, the energy used for lighting at a petroleum refinery facilities represent a small portion of the overall energy usage. However, there are opportunities for cost-effective energy efficiency improvements. Automated lighting controls that shut off lights when not needed may have payback periods of less than 2 years. Replacing T-12 lights with T-8 lights can reduce energy use by half, as can replacing mercury lights with metal halide or high pressure sodium lights. Substituting electronic ballasts for magnetic ballasts can reduce energy consumption by 12-25 percent (Worrell and Galitsky, 2005, 2008).

Lighting system improvements can be implemented at existing facilities and should be considered in the design of new construction.

5.0. GHG REDUCTION MEASURES BY SOURCE

5.1. Stationary Combustion Sources

5.1.1. Steam Generating Boilers

According to Worrell and Galitsky (2005), approximately 30 percent of onsite energy use at domestic refineries is used in the form of steam generated by boilers, cogeneration, or waste heat recovery from process units. The U.S. DOE estimated steam accounts for 38 percent of a refinery's energy needs (U.S. DOE, 2002). However, off-site purchases of steam represent only 3 to 5 percent of the total energy consumption at petroleum refineries nationwide (EIA, 2009). Given that steam accounts for 30 to 38 percent of a refinery's energy needs, it is evident that most refineries produce their own steam. As such, steam generation and distribution makes a significant contribution to a petroleum refinery's energy needs, and subsequently its on-site GHG emissions.

5.1.1.1. Systems Approach to Steam Generation

A thorough analysis of steam needs and energy recovery opportunities could be conducted to make the steam generation process as efficient as possible. For example, the analysis should assure that steam is not generated at pressures or volumes larger than what is needed. In those situations where the steam generation has limited adjustability, the excess energy in the steam should be recovered using a turbo expander or steam expansion turbine. Another option is to operate multiple boilers that are regulated according to steam demands. One refinery that implemented a program including scheduling of boilers on the basis of efficiency and minimizing losses in the turbines resulted in $5.4 million in energy savings (Worrell and Galitsky, 2005).

5.1.1.2. Boiler Feed Water Preparation

Boiler feed water is typically pre-treated to remove contaminates that foul the boiler. A refinery in Utah replaced a hot lime water softener with a reverse osmosis membrane treatment system to remove hardness and reduce alkalinity. Blowdown was reduced from 13.3 percent to 1.5 percent of steam produced. Additionally, reductions were seen in chemical usage, maintenance, and waste disposal costs. The initial investment of the membrane system was $350,000 and annual savings of $200,000 were realized (Worrell and Galitsky, 2005).

5.1.1.3. Improved Process Control

Boilers are operated with a certain amount of excess air to reduce emissions and for safety considerations. However, too much excess air may lead to inefficient combustion, and energy must be used to heat the excess air. Oxygen monitors and intake air flow monitors can be used to optimize the fuel/air mixture. Payback for such systems is typically about 0.6 years (Worrell and Galitsky, 2005).

5.1.1.4. Improved Insulation

The insulation of older boilers may be in poor condition, and the material itself may not insulate as well as newer materials. Replacing the insulation combined with improved controls can reduce energy requirements by 6-26 percent. Insulation on steam distribution systems should also be evaluated. Improving the insulation on the distribution pipes at existing facilities may reduce energy usage by 3-13 percent, with an average payback period of 1.1 years (Worrell and Galitsky, 2005).

5.1.1.5. Improved Maintenance

All boilers should be maintained according to a maintenance program. In particular, the burners and condensate return system should be properly adjusted and worn components replaced. Average energy savings of about 10 percent can be realized over a system without regular maintenance. Additionally, fouling on the fireside of the boiler and scaling on the waterside should be controlled (Worrell and Galitsky, 2005).

5.1.1.6. Recover Heat from Boiler Flue Gas

Flue gasses throughout the refinery may have sufficient heat content to make it economical to recover the heat. Typically, this is accomplished using an economizer to preheat the boiler feed water. One percent of fuel use can be saved for every 25 °C reduction in flue gas temperature. In some situations, the payback for installing an economizer is about 2 years (Worrell and Galitsky, 2005).

5.1.1.7. Recover Steam from Blowdown

The pressure drop during blowdown may produce substantial quantities of low grade steam that is suitable for space heating and feed water preheating. For boilers below 100 MMBtu/yr, fuel use can be reduced by about 1.3 percent, and payback may range from 1-2.7 years. A chemical plant installed a steam recover system to recover all of the blowdown steam from one process and realized energy savings of 2.8 percent (Worrell and Galitsky, 2005).

5.1.1.8. Reduce Standby Losses

It is common practice at most refineries to maintain at least one boiler on standby for emergency use. Steam production at standby can be virtually eliminated by modifying the burner, combustion air supply, and boiler feed water supply. Additionally, automatic control systems can reduce the time needed to reach full capacity of the boiler to a few minutes. These measures can reduce the energy consumption of the standby boiler by as much as 85 percent Worrell and Galitsky, 2005).

These measures were applied to a small 40 tonnes/hr steam boiler at an ammonia plant, resulting in energy savings of 54 TBtu/yr with a capital investment of about $270,000 (1999$). The payback period was 1.5 years (Worrell and Galitsky, 2005).

5.1.1.9. Improve and Maintain Steam Traps

Significant amounts of steam may be lost through malfunctioning steam traps. A maintenance plan that includes regular inspection and maintenance can reduce boiler energy usage by up to 10 percent (Worrell and Galitsky, 2005).

5.1.1.10. Install Steam Condensate Return Lines

Reuse of the steam condensate reduces the amount of feed water needed and reduces the amount of energy needed to produce steam since the condensate is preheated. The costs savings can justify the cost of the condensate return lines. Estimates of energy savings are as high as 10 percent, with a payback period of 1.1 years for facilities with no or insufficient condensate return systems (Worrell and Galitsky, 2005).

5.1.2. Process Heaters

5.1.2.1. Draft Control

Excessive combustion air reduces the efficiency of process heater burners. At one domestic refinery, a control system was installed on three CDU furnaces to maintain excess air at 1 percent rather than the previous 3-4 percent. Energy usage of the burners was reduced by 3-6 percent and nitrogen oxide (NO_x) emissions were reduced by 10-25 percent. The cost savings due to reduced energy requirements was $340,000. Regular maintenance of the draft air intake systems can reduce energy usage and may result in payback periods of about 2 months (Worrell and Galitsky, 2005). Draft control is applicable to new or existing process heaters, and is cost-effective for a wide range of process heaters (20 to 30 MMBtu/hr or greater).

5.1.2.2. Air Preheating

The flue gases of the furnace can be used to preheat the combustion air. Every 35 °F drop in exit flue gas temperature increases the thermal efficiency of the furnace by 1 percent. The resulting fuel savings can range from 8-18 percent, and may be economically attractive when the flue gas temperature is above 650 °F and the heater size is 50 MMBtu/hr or more. Payback periods are typically on the order of 2.5 years. One refinery in the United Kingdom installed a combustion air preheater on a vacuum distillation unit (VDU) and reduced energy costs by \$109,000/yr. The payback period was 2.2 years (Worrell and Galitsky, 2005). Air preheating would require natural draft system to be converted to a forced draft system requiring installation of fans, which would increase electricity consumption and typically increase NOx emissions. Consequently, several factors, including process heater size and draft type as well as secondary impacts, need to be considered retrofitting existing process heaters. Air preheating is often much more economical and effective when considered in the design of a new process heater.

5.1.3. Combined Heat and Power (CHP)

The large steam requirements for refining operations and the continuous operations make refineries excellent candidates for combined heat and power (CHP) generation. Refineries represent one of the largest industry sources of CHP today with 103 active CHP plants with an electric generation capacity of 14.6 gigawatts (ICF, 2010). Currently, about 60-70 percent of the 137 refineries operating at the beginning of 2010 use CHP (ICF International, 2010; EIA, 2009).

About 75 percent of the refinery CHP capacity comes from natural gas-fired combined cycle power plants consisting of large combustion turbines with heat recovery steam generators (HRSG) producing power and steam. A portion of the steam produced is used to generate more power in back pressure steam turbines. These plants meet the facility steam loads but often produce much more power than is needed by the facility itself, and, therefore, export power to the electric grid. The next most common type of CHP system is a combustion turbine with heat recovery. These systems make up about 11 percent of the existing refinery CHP capacity. Again, these systems are fueled mostly with natural gas, but internally generated fuels (i.e., refinery fuel gas) are also used. Most of the remaining system CHP capacity is boilers producing high pressure steam that run through a back-pressure steam turbine to produce power and lower pressure steam for process use. These systems generally do not use natural gas but, instead, are fired with a variety of internally generated fuels, waste fuels, and even coal.

While CHP systems are already in use at the majority of domestic refineries, there are significant remaining opportunities to add CHP-based on evaluation of steam requirements met by boilers and by CHP (Worrell and Galitsky, 2005). In addition, there are opportunities to repower existing CHP plants making them larger and more efficient by adding newer, more efficient combustion turbines and by converting existing simple cycle plants to combined cycle operation by adding steam turbines for additional power. Additionally, as refineries install flare gas recovery systems, they may need to install CHP systems to provide a productive source for utilizing the recovered fuel gas. There may be no direct CO_2 reductions at refineries from this technology, but indirect reductions from displacing grid power. The level of reduction is a function of the CO_2 intensity of the displaced external power production.

CHP systems require a fairly substantial investment (\$1,000-2,500/kilowatt (kW)); however, the economics of CHP operation at refineries is generally very attractive. One refinery installed a 34 megawatt (MW) cogeneration unit in 1990 that consisted of two gas turbines and two heat recovery steam boilers. All facility electricity needs are met by the unit, and occasionally excess electricity is exported to the grid. Cost savings resulting from the onsite production of electricity were about \$55,000/day. CHP can also be economical for small refineries. One study for an asphalt refinery showed that a 6.5 MW gas turbine CHP unit would reduce energy costs by $3.8 million/yr with a payback period of 2.5 years (Worrell and Galitsky, 2005).

5.1.4. Carbon Capture

The post-combustion technologies listed below are generally end-of-pipe measures. It should be noted that petroleum refineries emit CO_2 from a number of different process, and the exhaust stacks for these emission points are numerous and scattered across the facility. The consideration of CO_2 capture and control at a refinery would likely be limited to the larger CO_2 emitting stacks, such as the FCCU, the fluid coking unit, the hydrogen plant, and large boilers or process heaters.

5.1.4.1. Oxy-Combustion

Oxy-combustion is the process of burning a fuel in the presence of pure or nearly pure oxygen instead of air. Fuel requirements are reduced because there is no nitrogen component to be heated, and the resulting flue gas volumes are significantly reduced (Barker, 2009).

The process uses an air separation unit to remove the nitrogen component from air. The oxygen-rich stream is then fed to the combustion unit so the resulting exhaust gas contains a higher concentration of CO_2, as much as 80 percent. A portion of the exhaust stream is discharged to a CO_2 separation, purification, and compression facility. The higher concentration of CO_2 in the flue gas directly impacts size of the adsorber (or other separation technique), and the power requirements for CO_2 compression. This technology is still in the research stage. The Petroleum Environmental Research Forum (PERF) is focusing on large refinery combustion sources, particularly the FCCU and crude oil process heaters.

5.1.4.2. Post-Combustion Solvent Capture and Stripping

Post-combustion capture using solvent scrubbing, typically using monoethanolamine (MEA) as the solvent, is a commercially mature technology. Solvent scrubbing has been used in the chemical industry for separation of CO_2 in exhaust streams (Bosoaga, 2009).

5.1.4.3. Post-Combustion Membranes

Membrane technology may be used to separate or adsorb CO_2 in an exhaust stream. It has been estimated that 80 percent of the CO_2 could be captured using this technology. The captured CO_2 would then be purified and compressed for transport. Initial projections of specific costs range from $55-63/tonne CO_2 avoided for cement manufacturing. The current state of this technology is primarily the research stage, with industrial application at least 10 years away. Positive aspects of membrane systems include very low maintenance (no regeneration required) (ECRA, 2009).

5.2. Fuel Gas Systems and Flares

5.2.1. Fuel Gas Systems

Many process units at the refinery, particularly atmospheric crude oil distillation, catalytic cracking, catalytic hydrocracking, thermal cracking, and coking processes, produce fuel gas that is commonly recovered for use in process heaters and boilers throughout the refinery. Typically a compressor is needed to recover the fuel gas at the fuel gas producing unit. The fuel gas generally needs to be treated to remove H_2S using amine scrubber systems. The remainder of the fuel gas system consists of piping and mix drums to transport the fuel gas to the various combustion sources at the refinery. Rather than repeating the GHG reduction measures for each potential fuel gas producing units, the GHG reduction measures for the fuel gas system are summarized here.

5.2.1.1. Compressor Selection

Different types of compressors have different propensities to leak. Based on emission factors for natural gas compressors, reciprocating compressors generally have approximately one-half the fugitive emissions of centrifugal compressors (U.S. EPA, 1999). Rod packing (e.g., Static-Pac) can be used to reduce fugitive emissions from reciprocating compressors, and dry seal centrifugal compressors have lower emissions (i.e., are less likely to leak) than those with wet seals (U.S. EPA, 1999). Thus, the projected methane emissions from fuel gas compressors could be considered in the selection of the type of compressor and fugitive controls used.

5.2.1.2. Leak Detection and Repair (LDAR)

LDAR programs have been used to reduce emissions of volatile organic compounds (VOC) from petroleum refineries for years. However, CH_4 is not a VOC, so current regulations do not generally require LDAR for refinery fuel gas systems or other high CH_4-containing gas streams. Leaks can be detected using organic vapor analyzers or specially designed cameras. LDAR programs commonly achieve emission reduction efficiencies of 80 to 90 percent; however, CH_4 emissions from leaking equipment components is expected to have a minimal contribution to the refinery's total GHG emissions.

5.2.1.3. Selection of Fuel Gas Sulfur Scrubbing System

Hydrogen sulfide in fuel gas is commonly removed by amine scrubbing. The scrubbing solution is typically regenerated by heating the scrubbing solution in a stripping column, typically using steam. The regeneration process can use significant energy, and the energy intensity (impacting CO_2 emissions) of the different processes should be considered (in conjunction with the sulfur scrubbing efficiencies) in selecting scrubbing technology. Some fuel gas, such as fuel gas produced by coking units, contain a significant quantity of other reduced sulfur compounds, such as methyl mercaptan and carbon disulfide, that are not removed by conventional amine scrubbing. The impact of these other reduced sulfur compounds on the resulting sulfur dioxide (SO_2) emissions from process heaters and other fuel gas combustion devices using coker-produced fuel gas should be considered for both energy efficiency (for GHG emission reductions) and total sulfur removal efficiency (for SO_2 emission reductions). Alternatives to conventional amine scrubbing (which uses dimethylethylamine, DMEA), include the use of proprietary scrubbing systems, such as FLEXSORB®, Selexol®, and Rectisol®, as well as using a mixture of solvents as in the Sulfinol process, additional conversion of sulfur compounds to H_2S prior to scrubbing, or using a direct fuel gas scrubbing/sulfur recovery technology like LoCat® or caustic scrubbers.

CO_2 is also removed by amine scrubbing; however, this will not really impact the CO_2 emissions from the plant unless sulfur recovery occurs offsite because the CO_2 will be emitted either from the combustion unit receiving the fuel gas or from the sulfur recovery unit receiving the sour gas from the amine scrubbers. Therefore, the CO_2 scrubbing efficiency of the amine scrubbers is not important; however, some light hydrocarbons may also dissolve in the amine solution and subsequently sent to the sulfur recovery plant in the sour gas stream. Most hydrocarbons in the sour gas will eventually be oxidized in the sulfur recovery plant, so entrainment of hydrocarbons does lead to additional CO_2 emissions. Therefore, scrubbing systems could be evaluated based on their sulfur removal efficiency, energy efficiency, and ability to not entrain hydrocarbons. Note that higher sulfur removal efficiencies may have an energy penalty (*i.e.*, requiring more regeneration steam per pound of treated fuel gas), so a holistic analysis is needed when selecting the sulfur scrubbing system.

5.2.2. Flares

5.2.2.1. Flare Gas Recovery

Flaring can be reduced by installation of commercially available recovery systems, including recovery compressors and collection and storage tanks. Such systems have been installed at a number of domestic refineries. At one 65,000 bpd facility in Arkansas, two flare gas recovery systems were installed that reduced flaring almost completely. This facility will use flaring only in emergencies when the amount of flare gas exceeds the capacity of the recovery system. The recovered gas is compressed and used in the refinery's fuel system. The payback period for flare gas recovery systems may be as little as 1 year (Worrell and Galitsky, 2005). Similar flare gas recovery projects have been reported in the literature (John Zinc Co, 2006; Envirocomb Limited, 2006; Peterson *et al.*, 2007; U.S. DOE, 2005), reducing flaring by approximately 95 percent. Based on emission inventory presented by Lucas (2008), nationwide CO_2 emissions from flaring at petroleum refineries were estimated to be 5 million metric tons. Provided that the recovered fuel can off-set natural gas purchases, flare gas recovery is generally cost-effective for recovering routine flows of flare gas exceeding 20 MMBtu/hr (approximately 0.5 to 1-million scf per day, depending on heat content of flare gas). Based on these estimates, flare gas recovery could reduce nationwide CO_2 emissions from flares by 3-million metric tons. The cost-effectiveness of flare gas recovery is highly dependent on the heating value of the flare gas to be recovered and the price of natural gas. For refineries that may have excess fuel gas, a flare gas recovery system may also need to include a combined heat and power unit to productively use the recovered flare gas as described in Section 5.1.

5.2.2.2. Proper Flare Operation

Poor flare combustion efficiencies generally lead to higher methane emissions and therefore higher overall GHG emissions due to the higher global warming potential (GWP) of methane. Poor flare combustion efficiencies can occur at very low flare rates with high crosswinds, at very high flow rates (*i.e.*, high flare exit velocities), when flaring gas with low heat content, and excessive steam-to-gas mass flows. Installing flow meters and gas composition monitors on the flare gas lines and having automated steam rate controls allows for improved flare gas combustion control, and minimizes periods of poor flare combustion efficiencies.

5.2.2.3. Refrigerated Condensers for Process Unit Distillation Columns

For refineries that are rich in fuel gas, an alternative to a flare gas recovery system and CHP unit may be the use of a refrigerated condenser for distillation column overheads. Product recovery may be limited by the temperature of the distillation unit overhead condenser, causing more gas to be sent to the refinery fuel gas system and/or flare. The recovery temperature can be reduced by installing a waste heat driven refrigeration plant. A refinery in Colorado installed such a system in 1997 on a catalytic reforming unit distillation column and was able to recover 65,000 bbl/yr of LPG that was previously flared or used as a fuel. The payback of the system was about 1.5 years (Worrell and Galitsky, 2005).

5.3. Cracking Units

5.3.1. Catalytic Cracking Units

5.3.1.1. Power/Waste Heat Recovery

The most likely candidate for energy recovery at a refinery is the FCCU, although recovery may also be obtained from the hydrocracker and any other process that operates at elevated pressure or temperature. Most facilities currently employ a waste heat boiler and/or a power recovery turbine or turbo expander to recover energy from the FCCU catalyst regenerator exhaust. Existing energy recovery units should be evaluated for potential upgrading. One refinery replaced an older recovery turbine and saw a power savings of 22 MW and will export 4 MW to the power grid. Another facility replaced a turbo expander and realized a savings of 18 TBtu/yr (Worrell and Galitsky, 2005).

5.3.1.2. High-Efficiency Regenerators

High efficiency regenerators are specially designed to allow complete combustion of coke deposits without the need for a post-combustion device reducing auxiliary fuel combustion associated with a CO boiler.

5.3.1.3. Additional Considerations

Catalytic cracking units are significant fuel gas producers. As such, an FCCU can significantly alter the fuel gas balance of the refinery and may cause the refinery to be fuel gas rich (produce more fuel gas than it consumes) or increase the frequency of flare gas system over-pressurization to the flare. GHG measures for fuel gas systems could be considered. Flare gas recovery for the impacted flare(s) could also be considered. Also, an FCCU will have a process heater to heat the feed, so GHG reduction measures for process heaters may also need to be considered. Finally, as FCCUs are one of the largest single CO_2 emission sources at a refinery, carbon capture techniques (Section 5.1.4) could be considered.

5.3.2. Hydrocracking Units

5.3.2.1. Power/Waste Heat Recovery

For hydrocracker units, power can be recovered from the pressure difference between the reactor and fractionation stages. In 1993, one refinery in the Netherlands installed a 910 kW power recovery turbine to replace the throttle at its hydrocracker unit at a cost of $1.2 million

(1993$). The turbine produced about 7.3 million kilowatt hour per year (kWh/yr) and had a payback period of 2.5 years (Worrell and Galitsky, 2005).

5.3.2.2. Hydrogen Recovery

The hydrocracking unit is a significant consumer of hydrogen. Therefore, it is likely that a hydrocracking unit will significantly impact hydrogen production rates at the refinery (if the hydrogen production unit is captive to the refinery, i.e., under common ownership or control). The off-gas stream of the hydrocracker contains a significant amount of hydrogen, which is typically compressed, recovered, and recycled to the hydrocracking unit. When the recovery compressor fails or is taken off-line for maintenance, this high hydrogen gas stream is typically flared. A back-up recovery compressor could be considered for this high hydrogen stream. Although the flaring of hydrogen does not directly produce GHG, if natural gas is added to supplement the heating value of the flare gas, then flaring of the gas stream generates GHG. More importantly, the recovery of the hydrogen in this off-gas directly impacts the net quantity of new hydrogen that has to be produced for the unit. As hydrogen production has a large CO_2 intensity, continuous recovery of this high hydrogen gas stream can result in significant CO_2 emission reductions. At one Texas refinery, replacement of the hydrogen gas stream recovery compressor took 6 months, over which period approximately 7,000 tonnes of H_2 was flared, which corresponds to 63,000 to 70,000 tonnes of CO_2 emissions from additional hydrogen production. Considering the annualized capital cost of a back-up recovery compressor, the costs associated with the GHG emission reductions in this instance would be approximately $20 per tonne of CO_2 reduced.

5.3.2.3. Additional Considerations

Hydrocracking units produce fuel gas. As such, GHG measures for fuel gas systems are likely applicable for hydrocracking units. Additionally, flare gas recovery for the impacted flare(s) could be considered. The hydrocracking unit will have a process heater to heat the feed, so GHG reduction measures for process heaters may also need to be considered.

5.4. Coking Units

5.4.1. Fluid Coking Units

5.4.1.1. Power/Waste Heat Recovery

The fluid coking unit is an excellent candidate for energy recovery at a refinery. A CO boiler is used to combust the high CO off-gas from the fluid coking unit. Steam generation and/or a power recovery turbine or turbo expander could be used to recover energy from the CO boiler and its exhaust stream. Existing energy recovery units could be evaluated for potential upgrading.

5.4.1.2. Additional Considerations

Fluid coking units are significant fuel gas producers; GHG measures for fuel gas systems should be considered. Flare gas recovery for the impacted flare(s) could also be considered. The fluid coking unit will have a process heater to preheat the feed. Heat recovery systems could be considered for feed preheat; GHG reduction measures for process heaters may also need to be considered. Finally, as fluid coking units are one of the largest single CO_2 emission sources at a refinery, carbon capture techniques (Section 5.1.4) could be considered.

5.4.2. Flexicoking Units

Flexicoking coking units primarily produce a low-heating value fuel gas. Heat recovery from the produced gas stream should be used to preheat feed or to generate steam. The low-heating value fuel gas is typically combusted in specialized boilers and the GHG reduction measures for boilers could be reviewed. Also, flare gas recovery for the impacted flares and GHG reduction measures for process heaters may also need to be considered.

5.4.3. Delayed Coking Units

5.4.3.1. Steam Blowdown System

Delayed coking units use steam to purge and cool coke drums that have been filled with coke as the first step in the decoking process. A closed blowdown system for this steam purge controls both VOCs and methane. The steam to the blowdown system from a DCU will contain significant concentrations of methane and light VOCs. These systems could be enclosed to prevent fugitive emissions from the offgas or collected water streams. The noncondensibles from the blowdown system could be either recovered or directly sent to a combustion device, preferably a process heater or boiler rather than a flare to recover the energy value of the light hydrocarbons. Note that the sulfur content of this gas may prevent its direct combustion without treatment to remove sulfur.

As noted previously in Section 5.1.1.7 (regarding steam generating boilers), the blowdown system could be designed to operate at low pressures, so the DCU can continue to purge to the blowdown system rather than to atmosphere for extended periods. Also, a recovery unit to recycle hot blowdown system water for steam generation should be evaluated to improve the energy efficiency associated with the DCU's steam requirements.

5.4.3.2. Steam Vent

The DCU "steam vent" is potentially a significant emission source of both methane and VOCs. While not completely understood, the emissions from this vent are expected to increase based on the coke drum vessel pressure and the average temperature when the steam off-gas is first diverted to the atmosphere at (rather than to the blowdown system) at the end of the coke drum purge and cooling cycle. Generally, cycle times of 16 to 20 hours are needed to purge, cool, and drain the coke drum vessels, cut the coke out, and preheat the vessel prior to receiving feed. In efforts to increase throughput of the unit, reduced cycle times are used, but this generally requires depressurization of the coke drum at higher temperatures and pressures leading to higher emissions. While larger coke drums may have slightly higher emissions than smaller coke drums, the temperature of the coke drum when the drum is first vented to atmosphere will have a more significant impact on the volume of gas vented to the atmosphere than does the size (volume) of the coke drum. Cycle times of less than 16 hours are an indicator that the purging/quench cycles may be too short, leading to excessive and unnecessary VOC and CH_4 emissions. 40 CFR Part 60 subpart Ja requires new DCU to not vent to the atmosphere until a vessel pressure of 5 psig or less is reached. At this pressure, the equilibrium coke bed temperature should be approximately 230°F. However, as the vessel will be continuously purging to the blowdown system, the bed temperature may be significantly higher even though the pressure of the vessel is below 5 pounds per square inch gauge (psig) depending on the cycle time. A DCU could be designed to allow depressurization to very low pressures (*e.g.*, 2 psig) prior to having to go to atmosphere (which will impact the blowdown system design) to allow flexibility in operation. Analysis of the CH4 and VOC emissions at different temperatures and pressures could be conducted to determine operational parameters for the DCU depressurization/steam vent.

5.4.3.2. Additional Considerations

Delayed coking units are significant fuel gas producers. As such, GHG measures for fuel gas systems and flares could be considered. The fluid coking unit will have a process heater to preheat the feed. Heat recovery systems could be considered for feed preheat; GHG reduction measures for process heaters may also need to be considered.

5.5. Catalytic Reforming Units

The catalytic reforming unit is a net producer of hydrogen, so it can be considered as a means to produce hydrogen needed for other processes at the petroleum refineries; more detailed discussion of this is provided in Section 5.7. The reforming reaction is endothermic, so the catalytic reforming unit has large process heaters to maintain the reaction; GHG reduction measures for the process heaters could be considered. The catalytic reforming unit will also produce fuel gas so that GHG reduction measures for fuel gas systems and flares could be considered.

5.6. Sulfur Recovery Units

Nearly all refineries use the Claus-based sulfur recovery units, although some small refineries use LoCatTM system. There are, however, some variations on the traditional Claus system (*e.g.*, SuperClaus® and EuroClaus®) and a variety of different tail gas treatment units that are used in conjunction with the Claus sulfur recovery systems (*e.g.*, SCOT, Beavon/amine; Beavon/Stretford; Cansolv®, LoCat®, and Wellman-Lord). The energy and CO_2 intensities of these different systems could be evaluated (in conjunction with their sulfur recovery efficiencies) for sulfur recovery systems.

5.7. Hydrogen Production Units

Hydrotreating and hydrocracking units consume hydrogen. Hydrogen is produced as a by-product in catalytic reforming units. Hydrogen may also be produced specifically in captive or merchant hydrogen production units, which typically use steam methane reforming (SMR) techniques. Due to the importance of hydrogen for key processes and the interlinking of processes, a facility-wide hydrogen assessment could be performed to assess energy and GHG improvements that can be made. This assessment could include an assessment of whether additional catalytic reforming capacity can meet the hydrogen needs. Although both catalytic reforming and SMR are endothermic and require significant heat input, catalytic reformers produce high octane reformate (cyclic and aromatic hydrocarbons) rather than CO_2 as a result of the reforming reactions. Therefore, catalytic reforming provides a less CO_2-intensive means of producing hydrogen as compared to SMR hydrogen production. However, there is a limited quantity of naphtha and a limited need for reformate, so catalytic reforming may not be a viable option for meeting all of the hydrogen demands of the refinery.

If a hydrogen production unit is necessary, SMR technology appears to be the most effective means of producing additional hydrogen at this time. The following technologies could be considered for SMR hydrogen production units.

5.7.1. Combustion Air and Feed/Steam Preheat

Heat recovery systems can be used to preheat the feed/steam and combustion air temperature. If steam export needs to be minimized, an increase in the combustion air and feed/steam temperature through the convective section of the reformer is an option that can reduce fuel usage by 42 percent and steam export by 36 percent, and result in a total energy savings of 5 percent compared to a typical SMR (ARCADIS, 2008).

5.7.2. Cogeneration

Cogeneration of hydrogen and electricity can be a major enhancement of energy utilization and can be applied with SMR. Hot exhaust from a gas turbine is transferred to the reformer furnace. This hot exhaust at ~540 °C still contains ~13-percent oxygen and can serve as combustion air to the reformer. Since this stream is hot, fuel consumption in the furnace is reduced. The reformer convection section is also used as a HRSG in a cogeneration design. Steam raised in the convection section can be put through either a topping or condensing turbine for additional power generation. This technology is owned by Air Products and Technip, and has been applied at six hydrogen/cogeneration facilities for refineries (ARCADIS, 2008).

5.7.3. Hydrogen Purification

There are three main hydrogen purification processes. These are pressure-swing adsorption, membrane separation, and cryogenic separation. The selection of the purification method depends, to some extent, on the purity of the hydrogen produced. Pressure-swing adsorption provides the highest purity of hydrogen (99.9+ percent), but all of these purification methods can produce 95 percent or higher purity hydrogen stream. When lower purity (i.e., 95%) hydrogen gas is acceptable for the refinery applications, then any of the purification methods are technically viable. In such cases, the energy and CO_2 intensity of the various purification techniques could be considered. The purification technique also impacts the ease by which CO_2 recovery and capture can be used. See also the carbon capture techniques in Section 5.1.4.

5.8. Hydrotreating Units

A number of alternative hydrotreater designs are being developed to improve efficiency. New catalysts are being developed to increase sulfur removal, and reactors are being designed to integrate process steps. While many of these designs have not yet been proven in production, others such as oxidative desulfurization and the S Zorb process have been demonstrated at refineries. The design of both modifications and new facilities could consider the current state of the art (Worrell and Galitsky, 2005). Hydrotreaters consume hydrogen, so new hydrotreating units may also increase hydrogen production at the facility (see Section 5.7). Hydrotreaters also produce sour gas so the GHG reduction options discussed for sulfur scrubbing technologies (Section 5.2.1.3) and sulfur recovery units (Section 5.6) could be considered.

5.9. Crude Desalting and Distillation Units

Before entering the distillation tower, crude undergoes desalting at temperature ranging from 240 to 330 °F. Following desalting, crude enters a series of exchangers, known as preheat train to raise the temperature of the crude oil to approximately 500 °F. A direct-fired furnace is typically then used to heat the crude oil to between 650 and 750 °F before the crude oil is transferred to the flash zone of the tower. The crude oil furnaces are among the largest process heaters at the refinery; GHG reduction measures for these furnaces could be considered. Also, as the crude distillation unit employs among the largest process heaters at a refinery, carbon capture techniques (Section 5.1.4) could be considered. Additional GHG reduction measures are described below.

5.9.1. Desalter Design
Alternative designs for the desalter, such as multi-stage units and combinations of AC and DC fields, may increase efficiency and reduce energy consumption (Worrell and Galitsky, 2005).

5.9.2. Progressive Distillation Design
In the conventional scheme, all the crude feed is heated to a high temperature through the furnace prior to entering the atmospheric tower. Some lighter components of crude are superheated in the furnace, resulting in an irreversible energy waste. The progressive distillation process uses a series of distillation towers working at different temperatures (see Figure 6). The advantage of progressive distillation is that it avoids superheating of light fractions to temperatures higher than strictly necessary for their separation. The energy savings with progressive distillation has been reported to be approximately 30 percent (ARCADIS, 2008). Crude heaters account for approximately 25 percent of process combustion CO_2 emissions (Coburn, 2007); therefore, progressive distillation can reduce nationwide GHG emissions from petroleum refineries by almost 5 percent.

5.10. Storage Tanks

5.10.1. Vapor Recovery or Control for Unstabilized Crude Oil Tanks
Crude oil often contains methane and other light hydrocarbons that are dissolved in the crude oil because the crude oil is "stored" within the wells under pressure. When the crude oil is pumped from the wells and subsequently stored at atmospheric pressures, CH_4 and other light hydrocarbons are released from the crude oil and emitted from the atmospheric storage tanks. Most refineries receive crude oil that has been stored for several days to several weeks

at atmospheric pressures prior to receipt at the refinery. These stabilized crude oils have limited GHG emissions. If a refinery receives crude oil straight from a production well via pipeline without being stored for several days at atmospheric pressures, the crude oil may contain significant quantities of methane and light VOC. When this "unstabilized" crude oil is first stored at the refinery at atmospheric conditions, the methane and gaseous VOC will evolve from the crude oil. Common tank controls, such as floating roofs, are ineffective at reducing these emissions. If a refinery receives unstabilized crude oil, a fixed roof tank vented to a gas recovery system of control device could be considered to reduce the GHG (particularly CH4) emissions from these tanks.

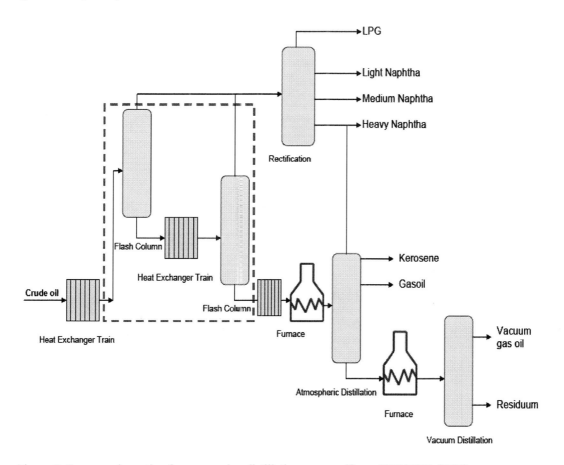

Figure 6. Process schematic of a progressive distillation process (from ARCADIS, 2008).

5.10.2. Heated Storage Tank Insulation

Some storage tanks are heated to control viscosity of the stored product. A study at a refinery found that insulating an 80,000 bbl storage tank that is heated to 225 °F could save $148,000 in energy costs (Worrell and Galitsky, 2005).

6.0. REFERENCES

[1]   ARCADIS. 2008. Air Pollution Control and Efficiency Improvement Measures for U.S. Refineries. Prepared for U.S. Environmental Protection Agency, Research Triangle Park, NC. Contract No. EP-C-04-023. August 8.

[2]   Barker, D.J., S.A. Turner, P.A. Napier-Moore, M. Clark, and J.E. Davison, 2009. "CO2 Capture in the Cement Industry," Energy Procedia, Vol. 1, pp. 87-94. http://www.sciencedirect.com/science?_ob=MImg&_imagekey=B984K-4W0SFYG-F-1&_cdi=59073&_user=10&_orig=browse&_coverDate=02%2F28%2F2009&_sk=999 989 998&view=c&wchp=dGLbVlzzSkWA&md5=12853bece66323782f9a 46335b4b 213c&ie=/sdarticle.pdf

[3]   Bosoaga, Adina, Ondrej Masek, and John E. Oakey. 2009. CO2 Capture Technologies for Cement Industry, Energy Procedia, Vol. 1, pp. 133-140. http://www.sciencedirect.com/science?_ob=MImg&_imagekey=B984K-4W0SFYG-N-1&_cdi=59073&_user=10&_orig=browse&_coverDate=02%2F28%2F2009&_sk=999 989 998&view=c&wchp=dGLzVlzzSkWA&md5=b9ba07fff56e3ad43069658cb689db9 e&ie=/sdarticle.pdf

[4]   Coburn J. 2007. Greenhouse Gas Industry Profile for the Petroleum Refining Industry. Prepared for U.S. Environmental Protection Agency, Washington, DC. Contract No. GS10F-0283K. June 11.

[5]   Ecofys. 2009. Methodology for the free allocation of emission allowances in the EU ETS post 2012. Sector report for the refinery industry. November.

[6]   ECRA (European Cement Research Academy). 2009. Development of State of the Art – Techniques in Cement Manufacturing: Trying to Look Ahead, June 4, 2009, Düsseldorf, Germany.), Cement Sustainability Initiative. http://www.wbcsdcement.org/pdf/technology/Technology%20papers.pdf

[7]   EIA (Energy Information Administration). 2006. Refinery Capacity Report 2006. Prepared by the Energy Information Administration, Washington, DC. June 15.

[8]   EIA (Energy Information Administration). 2009. Refinery Capacity Report 2008. Prepared by the Energy Information Administration, Washington, DC. June 25. See on-line HTML version of Table 12a: http://tonto.eia.doe.gov/dnav/pet/pet pnp capfuel dcu nus a.htm.

[9]   Envirocomb Limited. 2006. Zero flaring by flare gas recovery. Presented at the Gas Processors Association 1st Specialized Technical Session: Zero Flaring Seminar. Al-Khobar, Saudi Arabia, November 29. Available at: http://www.gpa-gcc-chapter.org/PDF/Zero%20Flaring%20By%20Flare%20Gas%20Recovery.pdf

[10]   Gary, J. H. and G.E. Handwerk. 1994. Petroleum Refining Technology and Economics. 3rd Edition. Marcel Dekker, Inc. New York, NY.

[11]   ICF International. 2010. CHP Installation Database, Maintained for U.S. DOE and Oak Ridge National Laboratory.

[12]   IPCC (International Panel on Climate Change). 2006. *2006 IPCC Guidelines for National Greenhouse Gas Inventories*. Prepared by the National Greenhouse Gas Inventories Programme. Edited by H.S. Eggleston, L. Buendia, K. Miwa, T. Ngara, and K. Tanabe. IGES: Japan.John Zinc Company. 2006. Flare gas recovery (FGR) to reduce plant flaring. Presented at the Gas Processors Association 1st Specialized Technical

Session: Zero Flaring Seminar. Al-Khobar, Saudi Arabia, November 29. Available at: http://www.gpa-gcc-chapter.org/PDF/Flare%20Gas%20Recovery%20White%20Paper.pdf.

[13] Lucas, Bob. 2008. Memorandum to Petroleum Refinery New Source Performance Standards (NSPS) Docket No. EPA-HQ-OAR-2007-0011 from Bob Lucas, USEPA regarding Documentation of Flare Recovery Impact Estimates. April 26.

[14] Peterson, J, N. Tuttle, H. Cooper, and C. Baukal. 2007. Minimize facility flaring. Hydrocarbon Processing. June. Pp. 111-115. Available at: http://www.johnzink.com/products/flares/pdfs/flare hydro proc june 2007.pdf.

[15] U.S. DOE (Department of Energy). 2002. Steam System Opportunity Assessment for the Pulp and Paper, Chemical Manufacturing, and Petroleum Refining Industries. U.S. Department of Energy, DOE/GO-102002-1640. October 2002.

[16] U.S. DOE (Department of Energy). 2005. Petroleum Best Practices Plant-wide Assessment Case Study. Valero: Houston refinery uses plant-wide assessment to develop an energy optimization and management system. DOE/GO-102005-2121. Available at: http://www1.eere.energy.gov/industry/bestpractices/pdfs/valero.pdf

[17] U.S. DOE (Department of Energy). 2007. Energy and Environmental Profile of the U.S. Petroleum Refining Industry. Prepared by Energetics, Inc., Columbia, MD. November 2007.

[18] U.S. EPA (Environmental Protection Agency). 1995. EPA Office of Compliance Sector Notebook Project: Profile of the Petroleum Refining Industry. EPA/310-R-95-013. Washington, DC. September 1995.

[19] U.S. EPA (Environmental Protection Agency). 1998. Petroleum Refineries-Background Information for Proposed Standards, Catalytic Cracking (Fluid and Other) Units, Catalytic Reforming Units, and Sulfur Recovery Units. EPA-453/R-98-003. Washington, DC: Government Printing Office.

[20] U.S. EPA (Environmental Protection Agency). 1999. U.S. Methane Emissions 1990–2020: Inventories, Projections, and Opportunities for Reductions. EPA 430-R-99-013. Section 3. Natural Gas Systems and Appendix III. Washington, DC. September 1999.

[21] U.S. EPA (Environmental Protection Agency). 2008. Technical Support Document for the Petroleum Refining Sector: Proposed Rule for Mandatory Reporting of Greenhouse Gases. Docket Item EPA-HQ-OAR-2008-0508-0025. Office of Air and Radiation. Office of Atmospheric Programs, Washington, DC. September 8.

[22] Worrell, Ernst and Christina Galitsky. 2005. Energy Efficiency Improvement and Cost Saving Opportunities for Petroleum Refineries (Report No. LBNL-56183). Ernest Orlando Lawrence Berkeley National Laboratory, Berkeley, CA. February 2005. http://www.energystar.gov/ia/business/industry/ES Petroleum Energy Guide.pdf

[23] Worrell, Ernst and Christina Galitsky. 2008. Energy Efficiency Improvement and Cost Saving Opportunities for Cement Making (Report No. LBNL-54036-Revision). Ernest Orlando Lawrence Berkeley National Laboratory, Berkeley, CA. March 2008. http://www.energystar.gov/ia/business/industry/LBNL-54036.pdf

In: Reducing Greenhouse Gas Emissions   ISBN: 978-1-61470-726-4
Editors: Diane B. McCreevey and Ellen L. Durkin   © 2011 Nova Science Publishers, Inc.

Chapter 3

AVAILABLE AND EMERGING TECHNOLOGIES FOR REDUCING GREENHOUSE GAS EMISSIONS FROM THE IRON AND STEEL INDUSTRY[*]

United States Environmental Protection Agency

ABBREVIATIONS AND ACRONYMS

°C	degrees Celsius
°F	degrees Fahrenheit
AC	alternating current
AISI	American Iron and Steel Institute
AIST	Association for Iron & Steel Technology
Al_2O_3	aluminum oxide
ANSI	American National Standards Institute
AOD	argon-oxygen decarburization
BACT	best available control technology
BFG	blast furnace gas
BOF	basic oxygen furnace
Btu	British thermal unit
CaO	calcium oxide
CCAP	Center for Clean Air Policy
cm	centimeter
CCS	carbon capture and sequestration
CDM	Clean Development Mechanism
CHP	combined heat and power
CFD	computational fluid dynamics
CH4	methane
CO	carbon monoxide

---

[*] This is an edited, reformatted and augmented version of the United States Environmental Protection Agency publication, dated October 2010.

CO_2	carbon dioxide
COG	coke oven gas
DC	direct current
DOE	U.S. Department of Energy
DRI	direct reduced iron
EAF	electric arc furnace
EMS	Energy Management Systems
EM Standards	Energy Management Standards
EPA	U.S. Environmental Protection Agency
FeO	iron oxide
ft	feet
ft^3	cubic foot
g/m^3	grams per cubic meter
GHG	greenhouse gas
GJ	gigajoule
gr/ft^3	grains per cubic foot
hr	hour
in	inch(es)
IPCC	Intergovernmental Panel on Climate Change
ISO	International Standards Organization
kg	kilogram
kVA	kilovolt amps
kWh	kilowatt hour
lbs	pounds
m	meter(s)
m^3	cubic meter
MMBtu	million British thermal units
MIT	Massachusetts Institute of Technology
MgO	magnesium oxide
mm	millimeters
NO_x	nitrogen oxide
PM	particulate matter
PM2.5	particulate matter with a diameter less than 2.5 micrometers (10^{-6} meters)
PSD	prevention of significant deterioration
PSH	paired straight hearth
R&D	research and development
scf	standard cubic feet
SiO_2	silicon dioxide
SOx	sulfur oxide
Tone	metric ton
Tpy	tons per year
UHP	ultra-high power
ULCOS	ultra-low CO2 steelmaking
VSD	variable-speed drive
yr	year

I. INTRODUCTION

This document is one of several white papers that summarize readily available information on control techniques and measures to mitigate greenhouse gas (GHG) emissions from specific industrial sectors. These white papers are solely intended to provide basic information on GHG control technologies and reduction measures in order to assist States and local air pollution control agencies, tribal authorities, and regulated entities in implementing technologies or measures to reduce GHGs under the Clean Air Act, particularly in permitting under the prevention of significant deterioration (PSD) program and the assessment of best available control technology (BACT). These white papers do not set policy, standards or otherwise establish any binding requirements; such requirements are contained in the applicable EPA regulations and approved state implementation plans.

Purpose of this Document

This document provides information on control techniques and measures that are available to mitigate greenhouse gas (GHG) emissions from the iron and steel manufacturing sector at this time. Because the primary GHG emitted by the iron and steel industry is carbon dioxide (CO_2), the control technologies and measures presented in this document focus on this pollutant. While a large number of available technologies are discussed here, this paper does not necessarily represent all potentially available technologies or measures that that may be considered for any given source for the purposes of reducing its GHG emissions. For example, controls that are applied to other industrial source categories with exhaust streams similar to the iron and steel manufacturing sector may be available through "technology transfer" or new technologies may be developed for use in this sector.

The information presented in this document does not represent U.S. EPA endorsement of any particular control strategy. As such, it should not be construed as EPA approval of a particular control technology or measure, or of the emissions reductions that could be achieved by a particular unit or source under review.

Description of the Iron and Steel Industry

The production of steel at an integrated iron and steel plant is accomplished using several interrelated processes. The major processes are (1) coke production, (2) sinter production, (3)iron production, (4) raw steel production, (5) ladle metallurgy, (6) continuous casting, (7) hot and cold rolling, and (8) finished product preparation. The operations for secondary steelmaking, where ferrous scrap is recycled by melting and refining in electric arc furnaces (EAFs) include (4)through (8) above. The interrelation of these operations is shown in a general flow diagram of the iron and steel industry in Figure 1.

The GHG emissions are generated as (1) process emissions, in which raw materials and combustion both may contribute to CO_2 emissions; (2) emissions from combustion sources alone; and (3) indirect emissions from consumption of electricity (primarily in EAFs and in finishing operations such as rolling mills at both integrated and EAF plants). The major process units at iron and steel facilities where raw materials, usually in combination with fuel combustion, contribute to the GHG emissions include the following:

- Sinter plant;
- Non-recovery coke oven battery combustion stack;
- Coke pushing;
- Basic oxygen furnace (BOF) exhaust; and
- EAF exhaust.

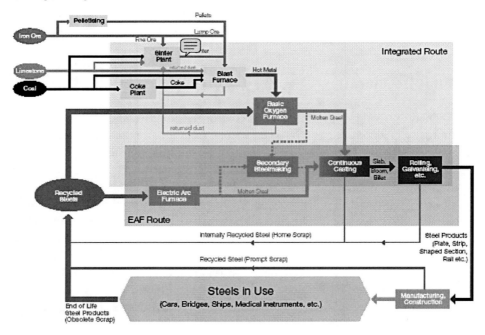

Source: International Iron and Steel Institute

Figure 1. Routes to steelmaking.

The primary combustion sources of GHGs include the following:

- By-product recovery coke oven battery combustion stack;
- Blast furnace stove;
- Boiler;
- Process heater;
- Reheat furnace;
- Flame-suppression system;
- Annealing furnace;
- Flare;
- Ladle reheater; and
- Other miscellaneous combustion sources.

For integrated steelmaking, the primary sources of GHG emissions are blast furnace stoves (43 percent), miscellaneous combustion sources burning natural gas and process gases (30 percent), other process units (15 percent), and indirect emissions from electricity usage (12 percent). For EAF steelmaking, the primary sources of GHG emissions include indirect emissions from electricity usage (50 percent), combustion of natural gas in miscellaneous combustion units (40 percent) and steel production in the EAF (10 percent). Additional information on the estimated GHG emissions is provided in Appendix A (Table A-4).

The following paragraphs provide brief descriptions of iron and steel processes. More detailed process descriptions, a list of plants and locations, and other industry information are provided in Appendix A.

Coke is the carbon product that is formed by the thermal distillation of coal at high temperatures in the absence of air in coke oven batteries. Coke is used in the blast furnace to provide a reducing atmosphere and is also a source of fuel. Most coke in the U.S. is produced in by-product recovery coke oven batteries, which recover tar, light oil, ammonia, and coke oven gas (COG) from the vapors generated in the ovens. Approximately one-third of the cleaned COG is used to fuel the coke ovens, and the balance is used in other combustion units at the steel plant. The four newest coke plants use non-recovery coke oven batteries that burn the by-products rather than recover them. The new non-recovery coke plants capture the waste heat from combustion to generate steam and electricity. The primary GHG emission point at coke plants is the battery's combustion stack.

Sintering is a process that recovers the raw material value of many waste materials generated at iron and steel plants that would otherwise be landfilled or stockpiled. Feed material to the sintering process includes ore fines, coke, reverts (including blast furnace dust, mill scale, and other by-products of steelmaking), recycled hot and cold fines from the sintering process, and trim materials (e.g., limestone, calcite fines, and other supplemental materials needed to produce a sinter product with prescribed chemistry and tonnage). The sinter feed materials are fused together by a flame fueled by natural gas and/or COG plus the ignition of coal and coke fines in the sinter feed. The product is a hard-fused material called sinter that is suitable for charging to the blast furnace. The primary emissions point of interest for the sinter plant is the stack that discharges the exhaust gases after gas cleaning. The CO_2 is formed from the fuel combustion (COG or natural gas) and from carbon in the feed materials, including limestone, coke fines, and other carbonaceous materials.

Iron is produced in *blast furnaces* by the reduction of iron-bearing materials with a hot gas. The large, refractory-lined furnace is charged through its top with iron ore pellets (taconite), sinter, flux (limestone and dolomite), and coke, which provides the fuel and forms a reducing atmosphere in the furnace. Many modern blast furnaces also inject pulverized coal or other sources of carbon to reduce the quantity of coke required. Iron oxides, coke, coal, and fluxes react with the heated blast air injected near the bottom of the furnace to form molten reduced iron, carbon monoxide (CO), and slag. The molten iron and slag collect in the hearth at the base of the furnace and are periodically removed from the furnace ("tapping"). The blast furnace gas (BFG) is collected at the top of the furnace and is recovered for use as fuel in the blast furnace stoves and other parts of the steel plant. The vast majority of GHGs (CO_2) are emitted from the blast furnaces' stove stacks where the combustion gases from the stoves are discharged. The carbon in the CO_2 exhaust comes mostly from the coke and coal used a fuel. A small amount of emissions may also occur from flares, leaks in the ductwork for conveying the gas, and from blast furnace emergency venting.

The **BOF** is a large, open-mouthed, pear-shaped vessel lined with a basic refractory material that refines molten iron from the blast furnace and ferrous scrap into steel by injecting a jet of high-purity oxygen to remove carbon as CO and CO2. The large quantities of CO produced by the reactions in the BOF are converted to CO2 by combustion at the mouth of the furnace in BOFs equipped with open hoods that draw in outside air or by flaring after gas cleaning in BOFs with tight-fitting closed hoods (called suppressed combustion). Final gas cleaning is performed by either venturi scrubbers or electrostatic precipitators for open hood BOFs; however, only venturi scrubbers are used on closed hood BOFs because of the explosion hazard from electrostatic precipitators if they were to be applied to the gas stream that is rich in CO. The major emission point for CO2 from the BOF is the furnace exhaust gas that is discharged through a stack after gas cleaning. The carbon in the CO2 exhaust comes mostly from the iron and scrap. Carbon may also be introduced into the BOF to a much smaller extent from fluxing materials and other process additives that are charged to the furnace.

Electric arc furnaces are used to produce carbon steels and alloy steels primarily by recycling ferrous scrap. Cylindrical refractory-lined EAFs are equipped with carbon electrodes that can be raised or lowered through the furnace roof. After ferrous scrap is charged, the electrodes are lowered and melting of the scrap begins when electrical energy is supplied to the carbon electrodes. Oxy-fuel burners and oxygen lances may also be used to supply chemical energy. Oxy-fuel burners, which burn natural gas and oxygen, use convection and flame radiation to transfer heat to the scrap metal. Some EAF plants, primarily the small specialty and stainless steel producers, use argon-oxygen decarburization (AOD) to further refine the molten steel from the EAF to produce low-carbon steel. In the AOD vessel, argon and oxygen are blown into the bottom of the vessel, and the carbon and oxygen react to form CO2 and CO, which are removed from the vessel. CO2 emissions from EAFs are generated primarily during the melting and refining processes, which remove carbon as CO and CO2 from the charge materials and carbon electrodes, and smaller quantities of CO2 are generated from the use of oxy-fuel burners by EAFs that are equipped with them. The emissions from the EAF are captured and sent to baghouses for removal of particulate matter (PM).

The steel produced by both BOFs and EAFs follow similar routes after the molten steel is poured from the furnace. The molten steel is transferred from ladle metallurgy to the continuous caster, which casts the steel into semi-finished shapes (e.g., slabs, blooms, billets, rounds, and other special sections). Steel from the continuous caster is processed in rolling mills to produce the final steel shapes that are sold by the steel mill. These shapes include coiled strips, rails, and other structural shapes, as well as sheets and bars. Because rolling mills consume electricity, they consequently contribute to indirect emissions of GHGs. The semi-finished products may be further processed by using many different steps, such as annealing, hot forming, cold rolling, heat treating (tempering), pickling, galvanizing, coating, or painting. Some of these steps require additional heating or reheating. The additional heating or reheating is accomplished using furnaces usually fired with natural gas. The furnaces are custom designed for the type of steel, the dimensions of the semi-finished steel pieces, and the desired temperature.

There are many different types of combustion processes at both integrated iron and steel and EAF steel facilities that are not directly related to the major production processes previously discussed. The EAF facilities burn natural gas almost exclusively in their combustion units, whereas integrated facilities burn a combination of fuels, including natural gas, COG, and BFG in their combustion units. The combustion units at both types of facilities include boilers, process heaters, flares, dryout heaters, and several types of furnaces. For example, soaking pits and reheat furnaces are used to raise the temperature of the steel until it is sufficiently hot to be plastic enough for economical reduction by rolling or forging. Annealing furnaces are used to heat the steel to relieve stresses formed through mechanical strain (hot or cold working) as well as stresses induced by rapid cooling (quenching). Annealing also softens the steel to improve machinability and formability. Ladle reheating uses natural gas to keep the ladle hot while waiting for molten steel.

II. ENERGY PROGRAMS AND MANAGEMENT SYSTEMS

Industrial energy efficiency can be greatly enhanced by effective management of the energy use of operations and processes. U.S. EPA's ENERGY STAR Program works with hundreds of U.S. manufacturers and has seen that companies and sites with stronger energy management programs gain greater improvements in energy efficiency than those that lack procedures and management practices focused on continuous improvement of energy performance.

Energy Management Systems (EMSs) provide a framework to manage energy and promote continuous improvement. The EMSs provide the structure for an energy program. The EMSs establish assessment, planning, and evaluation procedures which are critical for actually realizing and sustaining the potential energy efficiency gains of new technologies or operational changes.

The EMSs promote continuous improvement of energy efficiency through:

- Organizational practices and policies;
- Team development;
- Planning and evaluation;
- Tracking and measurement;
- Communication and employee engagement; and
- Evaluation and corrective measures.

For nearly 10 years, the U.S. EPA's ENERGY STAR Program has promoted an EMS approach. This approach, outlined in the graphic below, outlines the basic steps followed by most EMSs approaches (www.energystar.gov/guidelines).

In recent years, interest in EMS approaches has been growing. There are many reasons for the greater interest recently, which include recognition that a lack of management commitment is an important barrier to increasing energy efficiency. Further, lack of an effective energy team and energy efficiency program results in low implementation rates for new technologies or recommendations from energy assessments. Poor energy management practices that fail to monitor performance do not ensure that new technologies and operating procedures will achieve their potential in improving efficiency.

ENERGY STAR Guidelines for Energy Management.

Approaches to implementing EMSs vary. EPA's ENERGY STAR Guidelines for Energy Management are available for public use on the web and provide extensive guidance (see: www.energystar.gov/guidelines. Alternatively, Energy Management Standards (EM Standards) are available for purchase from American National Standards Institute (ANSI) as ANSI's Management System for Energy (ANSI MSE 2001:200),[1] and in the future from International Standards Organization (ISO), as ISO 50001.[2]

While EMSs can help organizations achieve greater savings through a focus on continuous improvement, they do not guarantee energy savings or carbon dioxide reductions alone. Combined with effective plant energy benchmarking and appropriate plant improvements, EMSs can help achieve greater savings.

There are a variety of factors to consider when contemplating requiring certification to an EM Standard established by a standards body such as ANSI or ISO. First, EMS standards are designed to be flexible. A user of the standard is able to define the scope and boundaries of the EMS so that single production lines, single processes, an entire plant or corporation could

---

[1] ANSI MSE 2001:200 can be found at http://www.mse2000.net/.

[2] Will be available from the ANSI webstore at http://webstore.ansi.org/default.aspx.

be certified. Achieving certification for the first time is not based on efficiency or savings (although re-certifications at a later time could be). Finally, cost is an important factor in the standardized approach. Internal personnel time commitments, external auditor and registry costs are high.

From a historical perspective, few companies have pursued certification according to the ANSI EM Standards to date. One reason for this is that the elements of an EMS can be applied without having to achieve certification, which adds additional costs. The ENERGY STAR Guidelines and associated resources are widely used and adopted partly because they are available in the public domain and do not involve certification.

Overall, a systems approach to energy management is an effective strategy for encouraging energy efficiency throughout a facility or corporation. The focus of energy management efforts are shifted from a "projects" to a "program" approach. There are multiple pathways available for the creation of EMSs with a wide range of associated costs (ENERGY STAR energy management resources are public while ANSI or ISO standardized approaches are costly). The effectiveness of an EMS is linked directly to the system's scope, goals, and monitoring and recordkeeping. Benchmarks are the most effective measure for demonstrating the system's achievements.

A. Sector-Specific Plant Energy Performance Benchmarks

Plant energy benchmarking is the process of comparing the energy performance of one site against itself over time or against the range of performance of the industry. Plant energy benchmarking is typically done at a whole-facility or site level in order to capture the synergies of different technologies, operating practices, and conditions.

Benchmarking enables companies to set informed and competitive goals for plant energy improvement. Benchmarking also helps companies prioritize where to invest to improve poorly performing systems while learning from the approaches used by top performing systems.

When benchmarking is conducted across an industrial sector, a benchmark can be established that defines best in class energy performance. EPA's ENERGY STAR Program has developed benchmarking tools that establish best-in-class for specific industrial sectors. These tools, known as Plant Energy Performance Indicators, are established for specific industrial sectors and available for free at www.energystar.gov/industrybenchmarkingtools. Using several basic plant-specific inputs, the Plant Energy Performance Indicators calculate a plant's energy performance providing a score from 0 to 100. The EPA defines the average plant within the industry nationally at the score of 50; energy-efficient plants score 75 or better. ENERGY STAR offers recognition for sites that score in the top quartile of energy efficiency for their sector using the Plant Energy Performance Indicators.

Another resource for policy-makers is the Center for Clean Air Policy (CCAP) and CCAP's publications that provide an international perspective on climate policy. The CCAP is performing research and providing policy support to help policy-makers around the world develop, promote, and implement innovative, market-based solutions to major climate, air quality, and energy problems that balance both environmental and economic interests. The mission statement of the CCAP is to significantly advance cost-effective and pragmatic air quality and climate policy through analysis, dialogue, and education to reach a broad range of policy-makers and stakeholders worldwide. (For more information, see http://www.ccap.org/.) In a study of global sectoral approaches (CCAP, 2010), CCAP investigated a transnational approach in which all countries face similar benchmarks, a sectoral Clean Development Mechanism (CDM) approach emphasizing carbon credits, and a bottom-up approach envisaging financial and technology assistance from advanced economies to support ambitious no-lose crediting baselines in developing countries. This study was supported by the Competitiveness and Innovation Framework Programme of the European Commission, and the study's objective was to help move beyond voluntary actions and facilitate participation by developing countries in international climate change actions.

B. Industry Energy Efficiency Initiatives

EPA's ENERGY STAR Program (www.energystar.gov/industry and the U.S. Department of Energy's (DOE) Industrial Technology Program (www.energy have led industry specific energy efficiency initiatives over the years. These programs have helped to create guidebooks of energy efficient technologies, profiles of industry energy use, and studies of future technologies. Some states have also led sector-specific energy efficiency initiatives. Resources from these programs can help to identify technologies that may help reduce CO2 emissions.

III. SUMMARY OF CONTROL MEASURES

Table 1 summarizes the GHG control measures presented in this document for integrated iron and steel plants, and Table 2 summarizes the measures for EAF steelmaking. Most of the information in the tables and the descriptions of the options in the next section were taken from a study by Worrell et al. (1999) that was conducted for EPA and DOE and subsequently updated (Worrell et al., 2009) in a study for EPA's Climate Protection Partnerships Division as part of the ENERGY STAR Program. The two tables include the emission reduction potential, energy savings, costs, and feasibility of each measure when such information was available. The following section provides more details and descriptions of the measures.

There are several important caveats associated with the estimated costs and savings presented in these tables that require caution in extrapolating the estimates in $/ton to specific iron and steel plants. For many of the measures, the costs and savings are based on the experience of a single plant or an individual application of the measure, or in some cases, best estimates based on the available information. The actual costs could be quite different when applying the measures to other individual plants because of the numerous site-specific differences among plants that affect costs. In addition, some measures may not be applicable to certain plants because of the process configuration, product type or quality constraints, or the fact that the measure or a similar one has already been applied. Some equipment modifications may incur significant retrofit costs that are not included in the estimates. The calculation of cost effectiveness introduces additional uncertainty because of the additional variability and uncertainty associated with potential reductions in GHG emissions. The choice of which measures might be the most appropriate to implement at a given facility should be based on a detailed analysis by the facility's engineering or energy manager to assess site-specific costs, savings, and emission reductions.

Table 1. Energy Efficiency Technologies and Measures Applied to Integrated Steel Production in the U.S. (Worrell et al., 1999, 2009)

Option	Emission Reduction (kg of CO_2/tone of product)	Fuel Savings (GJ/tone of product)	Electricity Savings (GJ/tonne of product)	Annual Operating Costs ($/tonne of product)	Retrofit Capital Costs ($/tonne of product)	Payback Time (years)
Iron Ore Preparation (Sintering)						
Sinter plant heat recovery	57.2	0.55	0.0	0.0	4.7	2.8
Emission optimized sintering						
Reduction of air leakage	2.0	0.0	0.0	0.0	0.14	1.3
Increasing bed depth	9.9	0.09	0.0	0.0	0.0	0.0
Improved process control	5.0	0.05	0.0	0.0	0.21	1.4
Use of waste fuels (e.g., lubricants) in sintering plant	19.5	0.18	0.0	0.0	0.29	0.5
Improve charging method						
Improve ignition oven efficiency						
Cokemaking						
Coal moisture control	6.7	0.30	0.0	0.0	76.6	> 50
Programmed heating	3.8	0.17	0.0	0.0	0.37	0.7
VSD COG compressor	0.12	0.0	0.0	0.0	0.47	21.2
Coke dry quenching	27.5	1.2	0.0	0.78	109.5	35.7
Additional use of COG						
Single chamber system						
Non-recovery coke ovens						
Ironmaking – Blast Furnace						
Pulverized coal injection to 130 kg/ton iron	47.0	0.77	0.0	-3.1	11.0	2.0
Pulverized coal injection to 225 kg/ton iron	34.7	0.57	0.0	-1.6	8.1	2.4
Injection of natural gas to 140 kg/ton iron	54.9	0.90	0.0	-3.1	7.8	1.3
Injection of oil						

Table 1. (Continued)

Option	Emission Reduction (kg of CO$_2$/tone of product)	Fuel Savings (GJ/tone of product)	Electricity Savings (GJ/tonne of product)	Annual Operating Costs ($/tonne of product)	Retrofit Capital Costs ($/tonne of product)	Payback Time (years)
Injection of COG and BOF gas						<1.0
Charging carbon composite agglomerates						
Top pressure recovery turbines (wet type)	17.6	0.0	0.11	0.0	31.3	29.8
Recovery of BFG	4.0	0.07	0.0	0.0	0.47	2.3
Hot-blast stove automation	22.6	0.37	0.0	0.0	0.47	0.4
Recuperator hot-blast stove	4.9	0.08	0.0	0.0	2.2	8.7
Improvement of combustion in hot stove						
Improved blast furnace control systems	24.4	0.40	0.0	0.0	0.56	0.4
Blast furnace gas recycling						
Slag heat recovery						
Steelmaking – Basic Oxygen Furnace (BOF)						
BOF gas plus sensible heat	46.0	0.92	0.0	0.0	34.4	11.9
VSD on ventilation fans	0.51	0.0	0.003	0.0	0.31	9.9
Improvement of process monitoring and control						
Programmed and efficient ladle heating						
Casting						
Efficient caster ladle/tundish heating	1.1	0.02	0.0	0.0	0.09	1.3
Near net shape casting – thin slab	728.8	3.5	0.64	-54.8	234.9	3.3
Near net shape casting – strip				25% less		
General Measures for Rolling Mills						
Energy efficient drives	1.6	0.0	0.01	0.0	0.30	3.2
Gate communicated turn-off inverters						

Option	Emission Reduction (kg of CO_2/tone of product)	Fuel Savings (GJ/tone of product)	Electricity Savings (GJ/tonne of product)	Annual Operating Costs ($/tonne of product)	Retrofit Capital Costs ($/tonne of product)	Payback Time (years)
Install lubrication system			0.016			
Hot Rolling						
Proper reheating temperature						
Avoiding overload of reheat furnaces						
Hot charging	30.2	0.60	0.0	-2.1	23.5	5.9
Process control in hot strip mill	15.1	0.30	0.0	0.0	1.1	1.2
Recuperative burners	35.2	0.70	0.0	0.0	3.9	1.8
Flameless burners	60%	60%				
Insulation of furnaces	8.0	0.16	0.0	0.0	15.6	31.0
Walking beam furnace			25%			
Controlling oxygen levels and VSDs on combustion air fans	16.6	0.33	0.0	0.0	0.79	0.8
Heat recovery to the product		50%		32%		
Waste heat recovery (cooling water)	1.9	0.03	0.0	0.11	1.3	> 50
Cold Rolling and Finishing						
Heat recovery on the annealing line	17.5	0.30	0.02	0.0	4.2	4.0
Reduced steam use (pickling line)	9.9	0.19	0.0	0.0	4.4	7.3
Automated monitoring and targeting system	35.3	0.0	0.21	0.0	1.7	0.8
Inter-electrode insulation in electrolytic pickling line						
Continuous annealing						
General						
Preventive maintenance	35.7	0.43	0.02	0.03	0.02	4.0
Energy monitoring and management system	9.5	0.11	0.01	0.0	0.23	0.5
Combined heat and power/cogeneration	82.1	0.03	0.35	0.0	22.7	6.1

Table 1. (Continued)

Option	Emission Reduction (kg of CO₂/tone of product)	Fuel Savings (GJ/tone of product)	Electricity Savings (GJ/tonne of product)	Annual Operating Costs ($/tonne of product)	Retrofit Capital Costs ($/tonne of product)	Payback Time (years)
High-efficiency motors						
VSD—flue gas control, pumps, and fans	1.5	0.0	0.02	0.0	2.0	10.7

Note: VSD = variable speed drive.

Table 2. Energy Efficiency Technologies and Measures Applied to Secondary Steel Production in the U.S. (Worrell et al., 1999, 2009)

Option	Emissions Reduction (kg CO₂/tonne of product)	Fuel Savings (GJ/tone of product)	Electricity Savings (GJ/tone of product)	Annual Operating Costs ($/tonne of product)	Retrofit Capital Costs ($/tonne of product)	Payback Time (years)
Steelmaking - Electric Arc Furnace						
Improved process control (neural network)	17.6	0.0	0.11	-1.6	1.5	0.5
Adjustable speed drives (ASDs)	10.0		0.05		2.0	2–3
Transformer efficiency—ultra-high power transformers	10.0	0.0	0.06	0.0	4.3	5.2
Bottom stirring/stirring gas injection	11.7	0.0	0.07	-3.1	0.94	0.2
Foamy slag practice	10.6	0.0	0.07	-2.8	15.6	4.2
Oxy-fuel burners	23.5	0.0	0.14	-6.2	7.5	0.9
Post-combustion of the flue gases						
DC arc furnace	52.9	0.0	0.32	-3.9	6.1	0.7
Scrap preheating—tunnel furnace (Consteel)	35.2	0.0	0.22	-3.0	7.8	1.3

Option	Emissions Reduction (kg CO_2/tonne of product)	Fuel Savings (GJ/tone of product)	Electricity Savings (GJ/tone of product)	Annual Operating Costs ($/tonne of product)	Retrofit Capital Costs ($/tonne of product)	Payback Time (years)
Scrap preheating, post-combustion—shaft furnace (Fuchs)	35.3	-0.70	0.43	-6.2	9.4	1.0
Engineered refractories			0.036			
Airtight operation			0.36			
Contiarc furnace			0.72			
Flue gas monitoring and control	8.8	0.0	0.05	0.0	3.1	4.3
Eccentric bottom tapping on existing furnace	8.8	0.0	0.05	0.0	5.0	6.8
Twin-shell DC with scrap preheating	11.1	0.0	0.07	-1.7	9.4	3.5
Casting						
Efficient caster ladle/tundish heating	1.1	0.02	0.0	0.0	0.09	1.3
Near net shape casting - thin slab	265.3	3.2	0.64	-54.8	234.8	3.3
Near net shape casting - strip				25% less		
Hot Rolling						
Proper reheating temperature						
Avoiding overload of reheat furnaces						
Energy efficient drives in the rolling mill	1.6	0.0	0.01	0.0	0.30	5.9
Process control in hot strip mill	15.1	0.30	0.0	0.0	1.1	1.2
Recuperative burners	35.2	0.70	0.0	0.0	3.9	1.8
Flameless burners	60%	60%				
Insulation of furnaces	8.1	0.16	0.0	0.0	15.7	31.0

Table 2. (Continued)

Option	Emissions Reduction (kg CO_2/tonne of product)	Fuel Savings (GJ/tone of product)	Electricity Savings (GJ/tone of product)	Annual Operating Costs ($/tonne of product)	Retrofit Capital Costs ($/tonne of product)	Payback Time (years)
Walking beam furnace			25%			
Controlling oxygen levels and VSDs on combustion air fans	16.6	0.33	0.0	0.0	0.79	0.8
Heat recovery to the product		50%		32%		
Waste heat recovery (cooling water)	1.9	0.03	0.0	0.11	1.3	> 50
General						
Preventive maintenance	15.0	0.09	0.05	0.03	0.02	
Energy monitoring and management systems	3.7	0.02	0.01	0.0	0.23	0.9

Note: DC = direct current; VSD = variable-speed drive.

Costs were taken primarily from Worrell et al. (1999) and adjusted from 1994 to 2008 dollars using the Chemical Engineering Plant Cost Index.[3] In addition, costs and energy savings are presented as "per tonne"[4] of product from the process (e.g., where "product" is steel from steelmaking furnaces, coke from coke plants, and sinter from sinter plants) Reductions in fuel consumption result in reductions of direct emissions of GHGs at the steel plant, and reductions in electricity usage result in reductions of indirect emissions (i.e., emissions from the power plant supplying the electricity). The annual operating costs that are represented in the tables do not include the energy savings from fuel or electricity. The value of the energy savings for fuel and electricity are very site-specific and depend upon many factors, such as the region of the country, special contract rates (e.g., based on quantity used, and for electricity, whether it is consumed during periods of peak demand), and changes in market price over time (e.g., fluctuations in the price of natural gas).

IV. ENERGY EFFICIENCY IMPROVEMENTS

The iron and steel industry is energy intensive; consequently, many of the options available to reduce GHG emissions involve improved energy efficiency. Current energy consumption is approximately 19 million British thermal units per ton of steel (MMBtu/ton) (22.1 GJ/tonne) for integrated mills and 5.0 MMBtu/ton (5.8 GJ/tonne) for EAFs. DOE estimates that a reduction of 5.1 MMBtu/ton (5.9 GJ/tonne) (27 percent) is possible for integrated mills (half from existing technologies and half from research and development [R&D]). A reduction of 2.7 MMBtu/ton (3.1 GJ/tonne) (53 percent) is possible for EAFs (two-thirds from existing technologies) (DOE, 2005).

The American Iron and Steel Institute (AISI) stated that, with existing technologies and best practices, improvements in blast furnace efficiency are possible through optimized blast injection technologies and better sensors and process controls. Other near-term opportunities noted by AISI include blast furnace coal injection modeling (to reduce energy losses in the cokemaking process) and optimizing processes through minimizing the generation of scrap and oxides (AISI, 2010).

The energy efficiency improvement options discussed by DOE (EPA, 2007a), listed below, are technically available but may not be economically viable in all situations:

- Preventive maintenance (0.21 MMBtu/ton (0.24 GJ/tonne) in potential energy reductions);
- Installation of energy monitoring and management systems for energy recovery and distribution between processes (0.06 MMBtu/ton) (0.07 GJ/tonne);
- Coal moisture control and dry quenching in the cokemaking process (0.22 MMBtu/ton) (0.26 GJ/tonne);

---

[3] The Chemical Engineering Plant Cost Index accounts for the changes in costs over time and is used to provide costs on a common year basis for comparisons. In this case, costs in 1994 dollars are multiplied by 1.56 to estimate the costs in 2008 dollars. The multiplier of 1.56 is the 2008 cost index (575.4) divided by the 1994 cost index (368.1).

[4] A metric tonne is a unit of mass equal to 1,000 kg (2,205 lb); conversely, 1 ton (2,000 lb) is equal to 0.907 metric tonnes, and used mostly in the U.S. The U.S. ton is sometimes called a "short" ton.

- For ironmaking (the most energy-intensive process), pulverized coal and natural gas injection, top pressure recovery turbines, hot-blast stove automation, and systems for improved blast furnace control (combined 1.34 MMBtu/ton) (1.56 GJ/tonne); and
- Casting/hot rolling energy efficiency opportunities include thin-slab casting with tunnel furnace (0.93 MMBtu/ton) (1.08 GJ/tonne).

Process-related opportunities noted by AISI (EPA, 2007a) for EAF steelmaking include the following:

- Improvements in process control, such as increased electrical energy transfer efficiency, reduced tap-to-tap times, and increased percentage of power-on time;
- Improved scrap preheating and charging practices; and
- Improved post-combustion practices.

A. Energy Efficiency Options for Integrated Iron and Steel Production

The following descriptions of mitigation options correspond to those in Tables 1 and 2 and are taken primarily from Worrell et al. (1999 and 2009). In many cases, the descriptions of measures were taken verbatim from these two research reports. Specific references are given when new or additional material was obtained from other sources. Costs and payback times are presented when available. All costs are presented in 2008 dollars. All references to "per ton" or to "per tonne" refer to tons or tonnes of product from the process that is being discussed.

1. Sintering

Sinter Plant Heat Recovery

Heat recovered from the sinter plant can be used to preheat the combustion air for the burners and to produce high-pressure steam, which can then be used in steam turbines to generate power. Various systems exist for new plants (e.g., Lurgi emission optimized sintering process), and existing plants can be retrofitted. Based on a retrofitted facility in The Netherlands, fuel savings were estimated to be 0.47 MMBtu/ton (0.55 GJ/tonne) of sinter, and increased electricity generation was estimated to be 1.4 kilowatt hour per ton (kWh/ton) (0.0056 GJ/tonne) of sinter. The payback time was estimated as 2.8 years. Emissions of nitrogen oxide (NO_x), sulfur oxide (SO_x), and PM are expected to be reduced. Capital costs are approximately $4.28 ton ($4.72/tonne) of sinter. Steam generation with sinter cooler gases using a waste heat boiler is common in Japan and was reported to recover 0.22 MMBtu/ton sinter (0.25 GJ/tonne).

Emission Optimized Sintering

This process for sinter plants was developed by Outokumpu Technology in the 1990s and can be retrofitted with minimal production interference. It reduces the substantial off-gas volume by 50–60 percent through housing the entire sinter strand, re-circulating off-gases, and using its CO content as an energy source to minimize off-gas volumes. The process reduces off-gas cleaning investment costs, saves energy in the form of coke, reduces operational costs, and significantly reduces NO_x, SO_x, CO, and CO_2 emissions.

Reduction of Air Leakage

Reducing air leakage from the sintering plant reduces fan power consumption by approximately 2.7-3.6 kWh/ton (0.011-0.014 GJ/tonne) of sinter. Costs of repairs to fix the leaks were estimated to be $0.13/ton ($0.14/tonne) of sinter capacity. Payback time was estimated as 1.3 years. (Improving fan efficiency is a potential energy saving option in other iron and steel processes as well as in other industrial sectors.)

Increasing Bed Depth

Increasing the bed depth in the sinter plant can lower fuel consumption, improve product quality, and increase productivity slightly. Fuel consumption may decrease by 0.6 pounds (lbs) of coke/ton of sinter per 0.4 inch (in.) bed thickness increase (0.3 kilogram [kg] of coke/tonne of sinter per 10 millimeters [mm] bed thickness increase). Electricity savings may be 0.05 kWh/ton (0.002 GJ/tonne) of sinter.

Improved Process Control

Based on general experience with industrial control and management systems, improved process controls may result in savings of 2–5 percent of energy use. Assuming a 2 percent savings, this would equate to a primary energy savings of approximately 0.04 MMBtu/ton (0.05 GJ/tonne) of sinter. Capital costs were estimated to be $0.19/ton ($0.21/tonne) of sinter. The payback time was estimated as 1.4 years.

Use of Waste Fuels in Sinter Plant

Waste materials with available caloric content (e.g., oils from the cold rolling mill) can be used as fuel and reduce the energy demand satisfied by the primary fuel. Estimates of the energy savings for this measure are difficult to make without knowing the quality and quantity of the waste material. Use of the waste material may be limited by permitted emissions limits because oils and other organics in the sinter feed increase emissions of organic compounds (including benzene, other volatile organic compounds dioxins, etc.). Based on data from European mills, the energy savings may amount to 0.15 MMBtu/ton (0.18 GJ/tonne) of sinter. The savings for this measure depend on the composition and quantity of lubricants and the installed gas clean-up system at the sinter plant. One plant reportedly developed a waste recovery and waste injection system, at a cost of about $25 million, to recycle 200,000 tons (180,000 tonnes) per year of various materials. Capital costs were estimated to be $0.26/ton ($0.29/tonne) of sinter. The payback time was estimated as 0.5 years.

Improve Charging Method

Limonite (brown iron ore) used as a raw material for sintering is inexpensive, but it decreases the productivity in the sintering process because it combines strongly with water and has a coarse particle size. These problems can be overcome by using an improved charging method. The system adopts a drum chute and a segregation slit wire. The purpose of the drum chute is to reduce the height difference (dropping difference) in material charging, while the segregation slit wire controls the particle size distribution. Specifically, because a constant particle size is maintained, the permeability of the sintering mixture is increased, resulting in improved sintering efficiency, and the material return ratio due to poor sintering is reduced. This system was developed by a Japanese steelmaker and has been introduced at all its plants in Japan. Productivity improvement amounts to 5 percent and energy consumption due to coke use decreases by 0.07 MMBtu/ton sinter (0.08 GJ/tonne sinter) compared to a conventional charging system.

Improve Ignition Oven Efficiency

A large fuel reduction can be achieved by improving the ignition oven efficiency. In order to reduce the fuel needed for ignition ovens, a heat-retention oven was removed from a conventional ignition oven with a large oven capacity, and an ignition oven with less capacity was introduced, in which the inner pressure of the ignition oven was regulated by controlling each windbox located immediately under the ignition oven. Moreover, a burner that can achieve rapid heating and uniform ignition in the pallet width direction has been developed and introduced to realize large fuel reductions. This burner consists of fuel exhaust nozzles located in the sintering floor width direction and a slit-like burner tile containing these fuel exhaust nozzles. The fuel supplied from the fuel exhaust nozzles reacts with the primary air inside the burner tile, then to the secondary air supplied to flame the outer periphery area. By using the slit-like burner tile, non-flamed places could be eliminated, and by controlling the ratio between the primary air and the secondary air, the length of the flame could be controlled to minimize the ignition energy. In this case, the ignition energy was reduced by approximately 30 percent.

Other Measures

Other measures include the use of higher quality iron ores, low iron oxide (FeO) content, replacing silicon dioxide (SiO_2) with magnesium oxide (MgO), reduction of the basicity of the sinter, and the use of coarse coke breeze.

2. Cokemaking

Coal Moisture Control

Waste heat from the COG can be used to dry the coal used for cokemaking, which may reduce the fuel consumption in the coke oven by approximately 0.26 MMBtu/ton (0.3 GJ/tonne). The cost of equipment to control coal moisture for a plant in Japan was $69.5/ton ($76.6/tonne) of steel. Application of the technique leads to a reduction of 0.11-0.18 MMBtu/ton coal (0.13- 0.21 GJ/tonne) in carbonization heat requirements, while the strength of the coke[5] is improved by approximately 1.7 percent and productivity by about 10 percent. The payback time was estimated as over 50 years.

---

[5] The physical strength of the coke is a critical parameter when the coke is charged to the blast furnace to prevent damage to the furnace if the charge materials (called the "burden") were to collapse and block the passageway for the gases and blast air that move through the charge materials.

Programmed Heating

The use of programmed heating instead of conventional constant heating of the coke ovens can help ensure optimization of the fuel gas supply to the ovens during the coking process. This measure can result in fuel savings of 10 percent, or approximately 0.15 MMBtu/ton (0.17 GJ/tonne) of coke. Capital costs for the computer control system were estimated to be $113,250/coke battery, or approximately $ 0.33/ton ($0.37/tonne) of coke capacity. The payback time was estimated as 0.7 years.

Variable-Speed Drive COG Compressors

Although COG is generated at low pressures and then pressurized for transport in the internal gas grid, the COG flows vary over time due to the coking reactions. The use of variablespeed–drive (VSD) COG compressors can reduce the energy required for compression of the low-pressure gas for transport. The VSDs help to compensate for variability in the gas flow due to coking reactions. One facility in The Netherlands installed a VSD system at a cost of $0.43/ton ($0.47/tonne) of coke and realized energy savings of 0.005-0.007 MMBtu/ton (0.006– 0 008 GJ/tonne) of coke. The payback time was estimated as 21 years.

Coke Dry Quenching

Dry quenching of the coke, in place of wet quenching, can be used to recover sensible heat that would otherwise be lost from the coke while reducing dust. The steam recovery rate with this equipment is about 0.5 MMBtu/ton (0.55 GJ/tonne) coke. In addition, Nippon Steel's performance record shows that the use of coke manufactured by dry quenching reduces the amount of coke consumption in the blast furnace by 0.24 MMBtu/ton (0.28 GJ/tonne)[6] molten iron. The payback time was estimated as 36 years. For new plants, the cost of the dry-quenching system was estimated to be $99.3/ton ($109.5/tonne) of coke. Retrofit costs depend strongly on the facility layout and can run as high as $118-152/MMBtu ($112-144/GJ) saved. Coke dry quenching has not been applied to any coke plants in the U.S.

Additional Use of Coke Oven Gas

Although COG is a low-Btu gas, approximately 40 percent of the COG is used as a fuel in coke ovens in the U.S. In most U.S. steel plants, the remaining COG is used to fuel equipment such as reheat furnaces and boilers that supply steam for electricity generation, turbine driven equipment such as pumps and fans, and for process heat. To the extent that any of the COG is flared at a facility, it could instead be used in combustion processes to offset the consumption of natural gas.

Single Chamber System

Single chamber system coking reactors (formerly called Jumbo Coke Reactors) are coke ovens with large coke oven volume and widths, between 17.7-33.5 in. (450-850 mm). The process includes the use of preheated coal. The reactors are separate process controlled modules with rigid, pressure stable, heating walls to absorb high coking pressure. This allows much thinner heating walls to be constructed, thus improving heat transfer and combustion, and greatly increasing the design flexibility of the plant. The high load bearing capacity of the

---

[6] Using net calorific value of 28,299 GJ/Gg coking coal.

side walls allows a greater range of coal bends to be charged, and the larger dimension ovens decrease the emissions of pollutants into the environment. The coal preheater increases coal bulk density, reduces the coking time, improves productivity and leads to increased coke strength. It is expected that these coke ovens are able to take the place of current multi-chamber batteries with walls of limited flexibility. Single chamber system coking reactors have an improvement in thermal efficiency from 38 percent to 70 percent, but the technology is currently under development.

Non-recovery Coke Ovens

In the non-recovery coking process, raw COG and other by-products released from the coking process are combusted within the oven, offering the potential for heat recovery and cogeneration of electricity. As the ovens operate under reduced pressure and at a temperature at which all potential pollutants break down into combustible compounds, this technique consumes all by-products, eliminating much of the potential for air emissions during the coking process and water pollution associated with the conventional byproduct recovery process. The process thus requires a different oven design from that traditionally used, resulting in a larger required area. A COG treatment plant and wastewater treatment plant are not needed.

When the waste gas exits into a waste heat recovery boiler, which converts excess heat into steam for power generation, the process is called heat recovery cokemaking. The four newest coke plants built in the U.S. have been the heat recovery type. In Haverhill, Ohio, a plant produces 500,000 tons (450,000 tonnes) of coke per year while producing 220 tons/hr (200 tonnes/hr) of steam, some of which is used in a nearby chemical plant and some used to generate electricity. Another plant in Granite City, Illinois, produces 650,000 tons (590,000 tonnes) of blast furnace coke per year and approximately 250 ton/hr (225 tonne/hr) of superheated steam.

3. Blast Furnace

Pulverized Coal Injection

Almost all integrated iron and steel plants have implemented pulverized coal injection at varying injection rates. Pulverized coal and natural gas injection replaces the use of coke, thereby reducing coke production and saving the large amount of energy consumed in cokemaking, reducing emissions from coke ovens, and reducing maintenance costs. However, increasing fuel injection requires energy for the oxygen and coal injection, electricity, and equipment to grind the coal. Some amount of coke is still used as support material in the blast furnace. In one application, the average coal injection rate into the blast furnace increased from approximately 4 lbs/ton (2 kg/tonne) of hot metal to approximately 260 lbs/ton (130 kg/tonne) of hot metal. The energy savings in the blast furnace due to coal injection have been calculated at 3.23 MMBtu/ton (3.76 GJ/tonne) coal injected. Fuel savings were estimated to be 0.66 MMBtu/ton (0.77 GJ/tonne) of hot metal, with capital costs of $9.92/ton ($10.94/tonne) of hot metal. Operating costs may decrease by $2.83/ton ($3.12/tonne) of hot metal. Investment costs for coal grinding equipment were estimated to be $45-50/ton ($50-55/tonne) of coal injected. The payback time is estimated as 2.0 to 2.4 years.

For large-volume blast furnaces, the coal injection rate may be increased to 450 lbs/ton (225 kg/tonne) of hot metal. Resulting fuel savings were estimated to be 0.49 MMBtu/ton

(0.57 GJ/tonne) of hot metal, with additional investment costs of $7.38/ton ($8.13/tonne) of hot metal. Operating costs may be reduced by $1.41/ton ($1.56/tonne) of hot metal.

Natural Gas Injection

Natural gas injection is typically applicable only to medium-sized furnaces having production rates of 1.4-2.5 million tons/yr (1.3–2.3 million tonnes/yr). Natural gas injection is an alternative to coal injection, and its selection depends on the price of natural gas versus coal. Replacement rates for natural gas may range from approximately 0.9-1.15 ton natural gas/ton coke (0.9–1.15 tonne of natural gas/tonne of coke). Estimated capital costs are $7.1/ton ($7.82/tonne) of hot metal. Estimated cost savings range from $3.6-4.5/ton ($4–5/tonne) of hot metal, and energy savings for a typical process were estimated to be 0.8 MMBtu/ton (0.9 GJ/tonne) of hot metal. Natural gas can be injected simultaneously with pulverized coal. It was reported that the rate by which natural gas injection can compensate for coal injection was a value of 6,400-16,000 ft^3/ton (200-500 m^3/tonne), depending on fuel composition and technological conditions. The payback time is estimated as 1.3 years.

Oil Injection

Heavy fuel oil or waste oil can also be injected (1 ton of oil replaces 1.2 tons of coke) (0.9 tonnes of oil replaces 1.1 tonnes of coke). With oxy-oil technology, the amount of oil injected can be increased by 100 percent to a level of 0.13 ton/ton hot metal (0.13 tonne/tonne hot metal). Like natural gas, oil contains hydrogen, leading to decreased CO_2 emissions.

Injection of COG and BOF Gas

Coke oven gas and BOF gas can also be injected into blast furnace. The maximum level for COG injection at the tuyère level is thought to be 0.1 ton COG/ton hot metal (0.1 tonne COG/tonne hot metal). The replacement rate of COG is about 1 ton of COG for 0.98 ton of coke (0.9 tonne of COG for 0.89 tonne of coke). This limit is set by the thermochemical conditions in the furnace. A compressor unit is required for COG injection resulting in an additional energy consumption of about 185 kWh/ton COG (0.73 GJ/tonne). Analysis indicates that injection of pulverized coal leads to higher energy effectiveness than that of COG. This technique was used in a U.S. Steel plant in Pittsburgh, Pennsylvania. Coke oven gas was used in the blast furnace to replace part of the natural gas that was injected in the furnace. It was found that it was more effective to use the COG in the blast furnace than to fuel boilers that feed steam turbines to generate electricity. The payback period was under 1 year.

Charging Carbon Composite Agglomerates

Carbon composite agglomerates are mixtures of fine iron ore (hematite, magnetite, iron-bearing dust and pre-reduced iron-bearing ore fines) and fine carbonaceous materials (fine coke, fine coal, charcoal, and char) with some binding agents added to the mixture in most cases.

These agglomerates were tested in operating blast furnaces and blast furnace simulators and were found to improve the energy efficiency of a blast furnace. Furthermore, the effective use of non-coking coal, and iron-bearing dust and sludge in steel works would reduce the amount of raw materials needed and promote resource recycling.

Top Pressure Recovery Turbines (Wet Type)

Top pressure recovery turbines are used to recover the pressure in the furnace. Although the differential pressure between the furnace and atmosphere is low, the large volume of gas may make recovery of the furnace pressure economical. The turbine may produce approximately 14- 36 kWh/ton (0.054-0.14 GJ/tonne) of hot metal. Although the top pressure in most U.S. furnaces is too low for recovery, future upgrades to furnaces may result in pressures high enough to allow economical recovery. (Upgrades occur every few years when the furnace is shutdown and relined. During these events, there is an opportunity to upgrade other equipment associated with the furnace operation.) Typical investment for the turbine is approximately $28.4/ton ($31.3/tonne) of hot metal. The payback time is estimated as 30 years.

Recovery of Blast Furnace Gas

Approximately 1.5 percent of the gas used in the blast furnace may be lost during charging, which could be recovered. A recovery system has been installed on a furnace in The Netherlands at a cost of $0.43/ton ($0.47/tonne) of hot metal. Energy savings have been estimated to be approximately 17 kWh/ton (0.066 GJ/tonne) of hot metal. The payback time is estimated as 2.3 years.

Hot-blast Stove Automation

This measure can reduce energy consumption of the stoves by running the operation more efficiently and closer to optimum conditions. Energy savings typically range between 5 and 12 percent, and may reach 17 percent. Typically, this may equate to 93 kWh/ton (0.037 GJ/tonne) of hot metal. The installation of a control system on a furnace in Belgium had a payback period of 2 months. Investment costs were assumed to be approximately $0.43/ton ($0.47/tonne) of hot metal. The payback time is estimated as 0.4 years. At the former ISPAT Island plant, the application of a model based controller for the optimal operation of blast stoves has led to 6-7 percent reductions in natural gas use and an improvement of operational consistency.

Recuperator Hot-Blast Stove

The hot-blast stove flue gases can be used to preheat the combustion air of the blast furnace. Various systems have been implemented, with fuel savings ranging from 20-21 kWh/ton (0.080–0.085 GJ/tonne) of hot metal at a cost of approximately $19-21/MMBtu ($18–$20/GJ) saved (equivalent to approximately $2.0/ton [$2.2/tonne] of hot metal). Preheating can lead to an energy saving of approximately 0.3 MMBtu/ton pig iron (0.35 GJ/tonne). An efficient hot-blast stove can run without the need for natural gas. For a specific medium-type waste heat recovery device (consisting of two heat exchangers), a recovery rate of sensible heat of 40-50 percent and a reduction in heat consumption of about 0.108 MMBtu/ton pig iron (0.126 GJ/tonne) produced has been reported. The payback time is estimated as 8.7 years.

Improvement of Combustion in Hot Stove

Improvement of combustion through more efficient burners and adaptation of combustion conditions (fuel/oxygen ratio) are estimated to lead to savings of 0.03 MMBtu/ton (0.04 GJ/tonne) pig iron.

Improved Blast Furnace Control Systems

Control systems have been developed in Europe and Japan to improve the control of the blast furnace. Estimated energy savings were approximately 0.34 MMBtu/ton (0.4 GJ/tonne) of hot metal. Capital costs were estimated to be approximately $0.5 million per blast furnace, or approximately $0.51/ton ($0.56/tonne) of hot metal. The implementation of a closed-loop blast furnace automation system at Voest Alpine (Linz, Austria) has resulted in a reduced coke consumption of approximately 0.46 ton/ton of hot metal (0.46 tonne/tonne of hot metal) in 2000, as well as reduced steam consumption by approximately 10.5 ton/hr (9.5 tonne/hr). The payback time is estimated as 0.4 years.

Blast Furnace Gas Recycling

Recirculation of the reducing gas components (CO and H_2) of the blast furnace gas formed in the furnace has been considered as an effective method to improve the blast furnace performance, enhance the utilization of carbon and hydrogen, and reduce the emission of carbon oxides. Previously, various recycling processes have been suggested, evaluated, or practically applied for different objectives. These processes are distinguished from each other by: (1) use with or without CO_2 removal, (2) use with or without preheating, and (3) the position of injection. This technology has not been commercially developed or deployed but it is the focus of intensive R&D in the Ultra-Low CO_2 Steelmaking (ULCOS) program.

Slag Heat Recovery

In modern blast furnaces, around 0.25-0.30 ton (0.23-0.27 tonne) liquid slag with a temperature of approximately 2,640 degrees Fahrenheit (°F) (1,450 degrees Celsius [°C]) is produced per ton pig iron. None of the current slag heat recovery systems have been applied commercially. This is due to the technical difficulties that would arise in the development of a safe, reliable, and energy efficient system that does not influence the slag quality.[7] One difficulty is that heat recovery from slag only becomes practical when the slag is granulated.[8] If such a technique were to be developed, associated estimated savings would be approximately 0.30 MMBtu/ton (0.35 GJ/tonne) pig iron.

4. Basic Oxygen Furnace

BOF Heat and Fuel Gas Recovery

The BOFs in the U.S. are either open-hood or closed-hood vessels (approximately 50 percent of each type). When oxygen is blown into the BOF to remove carbon to make steel, most of the carbon is removed as CO. In the open-hood BOF, large quantities of outside air are drawn into the BOF exhaust hood to burn the CO, and the exhaust gas reaches temperatures of 3,500°F (1,900°C). The exhaust gas from the open-hood BOF has no fuel value; however, both types of BOFs offer opportunities for heat recovery because of the high temperatures of the exhaust gas. Closed-hood BOFs suppress or prevent the intake of air, and

---

[7] Slag quality is important because it is a useful and valuable by-product, and over the past 25 years, all blast furnace slag produced in the U.S. has been used. These uses include aggregate for road bases, asphalt concrete, and concrete; production of mineral wool, cement, and glass; structural fill; and railroad ballast.

[8] Granulated blast-furnace slag is produced by quickly quenching (chilling) molten slag to produce a glassy, granular product. The most common process is quenching with water, but air or a combination of air and water may be used.

the resulting exhaust gas at 3,000°F (1,650°C) is much lower in volume than from an open hood. In addition, the closed-hood BOF generates an exhaust gas with fuel value from the CO. Newer BOFs are the closed-hood design, which has lower operating costs; however, no new BOFs have been installed in the U.S. in more than 30 years.

The closed-hood BOFs offer the best opportunity for both heat and fuel recovery. Although heat and fuel gas recovery from BOFs is very common in Japan and Western Europe, it is not implemented in the U.S., probably because the economics may be unattractive for retrofitting to old BOF shops. However, BOF gas and heat recovery is one of the most beneficial energy-saving process improvements for steelmaking. In steel plants in other countries, which use BOF gas, the predominant use is in the boiler plant, either directly or blended with BFG. BOF gas and BFG have also been used in gas turbine–combined cycle units, which are much more efficient in producing power than a conventional boiler and steam turbine generator set.

The energy recovery would reduce CO_2 generation from the use of natural gas and electricity by approximately 0.05 ton of CO_2/ton of steel (0.05 tonne of CO_2/tonne of steel), which would offset a significant portion of the unavoidable generation of CO_2 from steelmaking (estimated at 0.11 to 0.16 ton of CO_2/ton of steel) (0.11 to 0.16 tonne of CO_2/tonne of steel). Energy savings range from 0.46 to 0.79 MMBtu/ton (0.53 to 0.92 GJ/tonne). The capital cost of the recovery system is estimated at roughly $22/ton ($20/tonne) of steel (or approximately $66 million for an average BOF shop with a production rate of 3 million tons per year (2.7 million tonnes per year)). The payback period is estimated as 12 years.

Variable-Speed Drives on Ventilation Fans

The BOF process is a batch process, leading to widely varying volumes of flue gas. Thus, installing VSD can reduce energy consumption. At one facility, VSDs reduced energy consumption by 20 percent, or 0.82 kWh/ton (0.003 GJ/tonne) of crude steel. Investment costs are approximately $1.7 million , or approximately $0.28/ton ($0.31/tonne) of crude steel. At the Burns Harbor steelmaking facility, VSDs and equipment modifications reduced energy use at the BOF by about 50 percent and also reduced operation and maintenance expenses. The payback time was under 2 years.

Improvement of Process Monitoring and Control

Various types of monitoring systems make it possible to increase process control, which can lead to increased productivity and energy and cost savings. Examples of such systems are exhaust gas analysis systems, a contour sensing system, and simultaneous determination of steel/slag composition. The monitored data can also be used as input into models of the BOF process, which can help to improve understanding and to optimize the process. An example of process control system is a management oxygen system for oxygen supply to the BOF process. The total savings due to this system are estimated at 1.5 percent of the electricity used for oxygen production. The payback time is approximately 3 years (Worrell, 2010).

Programmed and Efficient Ladle Heating

The ladle of the BOF vessel is preheated with gas burners. Fuel consumption for preheating the ladle containing liquid steel is estimated at 0.017 MMBtu/ton (0.02 GJ/tonne) liquid steel. Heat losses can occur through the lack of lids and through radiation. The losses can be reduced by installing temperature controls, installing hoods, by efficient ladle management (reducing the need for preheating), using recuperative burners, and use of oxy-fuel burners.

Programmed ladle heating minimizes the quantity of fuel required to bring ladles up to steel handling temperatures. This may include the scheduling of ladle heating to ensure that ladles are not kept on heat for excessive periods as well as control of the combustion process. Furthermore, an efficient burner for ladle pre-heating makes sure that fuels are used efficiently.

JFE Steel Corporation's West Japan Works in Kurashiki improved the design of their ladle heating system by adopting a high-speed online heating apparatus and by developing a control system that combined this heating system with the control of the blowers in a BOF. In comparison with the heat balance before the introduction of the new process, the ratio of the amount of heat stored in a ladle refractory to the amount of heat input into a ladle improved by more than 10 times from 6.5 percent to 67.5 percent. The amount of heat stored in the ladle refractory at the time of receiving molten steel from the BOF became 6.3 times as large, which made it possible to reduce steel temperature at tapping by 16°F (9°C), thereby reducing the amount of carbon (coke) used to raise steel temperature in a converter by 16 percent.

5. Casting

Continuous casting replaced the operations of reheating cast ingots in soaking pits, and then rolling the ingots in roughing mills to produce slabs. Approximately 97 percent of the steel in the U.S. is produced via the continuous-casting method. The energy savings over ingot casting is approximately 2.46 MMBtu/ton (2.86 GJ/tonne) of steel cast with cost savings of approximately $28/ton ($31/tonne) of steel. One estimate placed investment costs at $63/ton ($69/tonne) of steel.

Efficient Ladle Preheating and Tundish Heating

The ladle of the caster is preheated with gas burners with a fuel consumption estimated at 0.02 MMBtu/ton (0.02 GJ/tonne) liquid steel. Heat losses can occur through lack of lids and through radiation. The losses can be reduced by installing temperature controls, installing hoods, by efficient ladle management (reducing the need for preheating), using recuperative burners, and using oxy-fuel burners.

Tundishes are heated to reduce the heat loss of the molten steel, to avoid bubbles in the first slab at the beginning of the casting sequence, and to avoid degeneration of the refractory due to thermal shocks. Combustion-heated tundishes on average are only 20 percent efficient. Although earlier tundishes heated through electrical induction failed to generate enough heat to be effective in the manufacturing process, new methods have been developed to improve heating capacity. Tundishes heated by electrical induction have the potential to reach efficiency levels of 98 percent; however, the use of electricity may result in indirect energy losses in power generation. Energy savings can also be attained by refraining from heating the tundish. Practices at a plant in Brazil have shown that the use of a cold tundish is operationally feasible and that it brings with it several main benefits: a 70 percent reduction in the time for machine return after interruptions at the beginning of the cycle, a 78 percent reduction of natural gas consumption, a 90 percent increase of the lifetime of the tundish lids, and improvement of the working conditions on the casting platform due to heat and noise reductions. (Percentages are expressed on a per process unit basis.) The practice was not found to have any influence on the quality of the product.

At North Star Steel, Iowa, it was estimated that the installation of recuperators for the ladle and tundish heating system would result in fuel savings of 28 percent at the ladle heaters and 26 percent at the tundish dryer. Payback periods were estimated to be 1.2 years and 10 years, respectively. While a tundish heater-dryer (capital cost $45,000) annually saves approximately 1,000 MMBtu (1,050 GJ) of natural gas, ladle heaters (capital cost $70,000) save 13,500 MMBtu (14,000 GJ) of natural gas per year. Although general estimates of the fuel savings are difficult to make, one estimate placed potential energy savings at 50 percent, or approximately 0.017 MMBtu/ton (0.02 GJ/tonne) of crude steel. Costs were estimated to be $0.08/ton ($0.09/tonne) of product, assuming a gas price of $3.0/MMBtu ($2.8/GJ), giving a payback period of 1.3 years.

Near Net Shape Casting

Near net shape casting is a process of casting metal to a form close to that required for the finished product. This means less machining is required to finish the part. Near net shape casting integrates the casting and hot rolling of steel into one process step, thereby reducing the need to reheat the steel before rolling it. Several production processes have been developed for near net shape casting, most notably thin slab casting and strip casting. Thin slab casting and strip casting are both forms of continuous casting.

In the case of thin slab casting, the steel is cast directly to slabs with a thickness between 1.2 and 2.4 in (30 and 60 mm) instead of slabs with a thickness of 4.72-11.8 in (120-300 mm). Thin slab casting has been a success in flat product mini-mills in the U.S., and this technology may be a future opportunity for a few more plants that produce thin slabs of steel. Thin slab casting is estimated to reduce energy consumption by 4.2 MMBtu/ton (4.9 GJ/tonne) of crude steel with a payback time of 3.3 years. Investment costs for a large-scale plant were estimated to range from $213/ton ($234.9/tonne) of product with a resultant cost savings of approximately $28/ton ($31/tonne) of crude steel. Another study indicated that thin slab casting with a tunnel furnace offered an energy savings of 0.93 MMBtu/ton (1.08 GJ/tonne) of steel cast.

In strip casting, the steel is cast between two rolls, producing directly a strip of around 0.12 in (3 mm) thickness. Three commercial technologies have emerged in which the steel is cast between two water-cooled casting rolls, which results in very rapid cooling and high production speeds. The major advantage of strip casting is the large reduction in capital costs, due to the high productivity and integration of several production steps. The technology was first applied to stainless steel, and two plants have demonstrated strip casting of carbon steel. One commercial strip casting technology is Castrip®, which was constructed at Nucor's Crawfordsville, Indiana plant in 2002, and since that time the plant has produced Ultra-Thin Cast Strip products. Nucor has also commenced construction of its second strip casting plant in Blytheville, Arkansas. Compared to thick slab casting (hot rolling, pickling and, cold rolling), thin slab casting saves approximately 0.9 MMBtu/ton (1 GJ/tonne). In turn, compared to thin slab casting, the Castrip® process saves approximately another 0.9 MMBtu/ton (1 GJ/tonne). Other strip casting technologies include Eurostrip (developed by a consortium of ThyssenKrupp Steel, Arcelor, and Voest Alpine Industries) and Nippon/Mitsubishi.

This strip casting technology leads to considerable capital cost savings and energy savings. It may also lead to indirect energy savings due to reduced material losses. Operations and maintenance costs are also expected to drop by 20-25 percent, although this will depend strongly on the lifetime of the refractory on the rollers used in the caster and local circumstances. Energy consumption of a strip caster is significantly less than that for continuous casting, with an estimated fuel use of 0.04 MMBtu/ton (0.05 GJ/tonne) and electricity use of 39 kWh/ton (0.15 GJ/tonne). The savings over traditional thick slab continuous casting include 80-140 kWh/ton (0.32-0.55 GJ/tonne) for electricity and 1.0-1.3 MMBtu/ton (1.2-1.5 GJ/tonne) for fuel (Worrell, 2010).

A new development of thin slab casting and direct rolling is "Endless Strip Production." Installation of this technology was started in 2008 at a plant in Cremona, Italy. The specific energy consumption should be 40 percent lower than that needed for a traditional rolling mill. For thin gauges, the suppression of the cold rolling and annealing cycle will allow energy savings of 60 percent with regard to the traditional cycle. Processing costs are characterized by lower energy consumption, lower costs for consumables (e.g., mould, rolling cylinders) and improved liquid steel yield (up to 93 percent).

6. General Measures for Rolling Mills

The semi-finished steel products from the casting operations are further processed to produce finished steel products in a series of shaping and finishing operations. This section presents some energy efficiency measures that are applicable to both the hot rolling and cold rolling processes, and the following sections discuss energy efficiency measures that are specific to each. Mechanical forces for cold rolling will create much more force and energy needs, while hot rolling happens much faster with less forces; however, there are significant energy costs to heat the metal to near eutectic temperatures.

Energy Efficient Drives

High-efficiency alternating current (AC) motors can save 1 or 2 percent of the electricity consumption of conventional AC drives. Based on an electricity demand of 181 kWh/ton (0.072 GJ/tonne) of rolled steel, the electricity savings were estimated to be 3.6 kWh/ton (0.014 GJ/tonne) of hot rolled steel. The additional cost of the high efficiency drives was estimated to be approximately $0.27/ton ($0.30/tonne) of hot rolled steel. The payback time is estimated as 3.2 years.

Gate Communicated Turn-Off Inverters

Drive units for main equipment such as rolling mills in steel plants use variable-speed AC operation. As switching devices for large-capacity inverted drives, Gate Turn-Off thyristors have been widely used. However, a Gate Communicated Turn-Off thyristor can be used instead of a Gate Turn-Off thyristor to decrease switching losses. Compared with this Gate Turn-Off the Gate Communicated Turn-Off inverter has higher system efficiency, not only at rated-load operation, but also at light-load operation and reduces energy loss. The Gate Communicated Turn-Off inverters are typically used to drive steel rolling mills and are being adopted in every area of steel mills from high-speed wire rolling to low-speed cold rolling. Moreover, they are applicable as energy-saving drive units for large-capacity fans, pumps, and compressors.

Install Lubrication System

High roll loads of steel lead to increased roll wear and high energy consumption. In addition, specific combinations of rolling loads and speeds can cause stands to vibrate which leads to a special type of roll banding and increased wear of the equipment. These problems can be solved by installing a lubrication system as was done at an EKO Stahl hot rolling mill. Energy savings due to this project were estimated to be about 4 kWh/ton (0.016 GJ/tonne), which is a savings of about $0.28/ton ($0.31/tonne).

7. Hot Rolling

In any hot rolling operation, the reheating furnace is a critical factor to determine end-product quality, as well as to total costs of the operation. Energy use in a reheating furnace depends on production factors (e.g., stock, steel type), operational factors (e.g., scheduling), and design features. Savings may be achieved through optimized processes and by upgrading existing furnaces. The upgrade of a reheat furnace of North Star Steel (Iowa) led to significant fuel, energy cost and labor savings together with savings due to the reduction of scrap use while furnace refractory life and product quality improved.

Proper Reheating Temperature

In choosing the heating temperature for semi-finished products prior to rolling, an attempt should be made to obtain a fine-grained structure in the metal along with the requisite mechanical properties in the rolled product. The heating operation should also ensure dissolution of the inclusions in the metal in the absence of excessive grain growth. A reduction of the heating temperature by 212°F (100°C) decreases unit fuel consumption by 9 to 10 percent. However, lowering the heating temperature will increase the rolling forces and moments, and hence increase the load on the electric drive motors, i.e., it will have the overall effect of increasing the mechanical and electrical loads on the main components of the mill, thereby increasing energy consumption and wear of the mill equipment. As a result, under certain conditions total unit energy consumption may not decrease with a decrease in heating temperature (even without allowance for the losses associated with electric power generation). The heating temperature should therefore be carefully considered using a systems approach.

Avoiding Overload of Reheat Furnaces

Overloading a furnace can lead to excessive stack temperatures. To get the proper rate of heat transfer, combustion gases must remain in the heating chamber for the right amount of time. The natural tendency of an overloaded furnace is to run colder than optimal, unless the temperature is set artificially high. This causes the burners to operate at higher than normal firing rates, which increases combustion gas volumes. The higher gas flow rates and shorter time that the gas remains in the furnace causes poor heat transfer, resulting in higher temperatures of the flue gases. The increased volumes of higher temperature flue gases lead to sharply increased heat losses. Overly ambitious production goals might be met, but at the cost of excessive fuel consumption. The overload problem may be corrected by improving heat transfer or not operating in this mode to achieve ambitious production goals.

Hot Charging

Hot charging is the process of heating slabs prior to charging them into the reheating furnace of the hot mill. The higher the preheat temperature, the greater the energy savings in the hot mill furnace. The layout of the plant will affect the feasibility of hot charging because the caster and reheating furnace should be located in proximity to one another to avoid a long, hot connection between the two. Although actual savings will be highly plant dependent, one estimate of the potential energy savings was as much as 0.05 MMBtu/ton (0.06 GJ/tonne) of hot charged steel. Investment costs were estimated to be approximately $21.3/ton ($23.5/tonne) of hot rolled steel, with annual cost savings of up to $1.04/ton ($1.15/tonne) of hot charged steel and a payback time of 5.9 years.

Process Control in Hot Strip Mill

Improved process control of the hot strip mill may lead to indirect energy savings through reduced product rejects, improved productivity, and reduced down time. This measure includes controlling oxygen levels and VSDs on combustion air fans, which both help to control the oxygen level, and hence optimize the combustion in the furnace, especially as the load of the furnace may vary over time. The savings depend on the load factor of the furnace and control strategies applied. A system installed at ArcelorMittal's Sidmar plant (Belgium) reduced the share of rejects from 1.5 to 0.2 percent and reduced the downtime from more than 50 percent of the time to 6 percent. Estimated energy savings based on reduced rejects was 9 percent of fuel use, or approximately 0.26 MMBtu/ton (0.3 GJ/tonne) of product. The investment costs for one plant in Belgium was $3.6 million for a hot strip mill with a capacity of 3.1 million tons (2.8 million tonnes), or approximately $1.20/ton ($1.29/tonne) of product. The payback time is estimated as 1.2 years.

Recuperative Burners

Application of recuperative or regenerative burners can substantially reduce energy consumption. A recuperator is a gas-to-gas heat exchanger placed on the stack of the furnace. There are numerous designs, but all rely on tubes or plates to transfer heat from the outgoing exhaust gas to the incoming combustion air, while keeping the two streams from mixing. Recuperative burners use the heat from the exhaust gas to preheat the combustion air. Recuperative burners can reduce fuel consumption by 10 to 20 percent compared to furnaces without heat recovery.

Since modern recuperative or regenerative burner systems can have significantly higher efficiencies than older systems, savings can also be attained by replacement of old or aging recuperative or regenerative burners. Newer designs can also have lower NO_x emissions; consequently, the evaluation of recuperative or regenerative burner systems should include an assessment of the impact on NO_x emissions. Replacement of the recuperator by a newer model can result in substantial savings as is illustrated by an example at North Star Steel (Iowa). Recuperator replacement at this plant was estimated to achieve fuel savings of 9 percent with an expected payback period of 6 months. Another example in Japan shows that a newer model continuous slab reheating furnace can reduce energy consumption by 25 percent in comparison to an older furnaces recovering waste heat with a recuperator.

Recuperative burners in the reheating furnace can reduce energy consumption by as much as 30 percent. Although actual savings will be highly facility-specific, one estimate placed energy savings at approximately 0.6 MMBtu/ton (0.7 GJ/tonne) of product, with an investment cost of approximately \$3.5/ton (\$3.9/tonne) of product. The payback time is estimated as 1.8 years.

Flameless Burners

A widely used technique to enhance furnace efficiency is extensive air preheating, but the drawback is a parallel increase of NO_x emissions. Another technique is the use of flameless burners. Flameless air-fuel combustion uses air as oxidizer, while flameless oxy-fuel uses commercial oxygen as an oxidant. This technology carries out combustion under diluted oxygen conditions using internal flue gas recirculation and the flame becomes invisible. Flameless oxyfuel gives high thermal efficiency, higher levels of heat flux, and reduced fuel consumption compared to conventional oxy-fuel. These benefits are combined with low NO_x emissions and better thermal uniformity. Since 2003, more than 30 furnaces within the U.S. steel industry have been equipped with flameless oxy-fuel combustion.

ArcelorMittal recently received the Association for Iron & Steel Technology (AIST) 2009 Energy Achievement Award for its work to implement a flameless oxy-fuel operation on its rotary-hearth steel-reheat furnace. ArcelorMittal realized a 60 percent reduction in the furnace's total fuel consumption compared to the original air-fuel operation. The technology also reduced the furnace's annual NO_x emissions output by 92 percent and annual CO_2 emissions by up to 60 percent below the prior air-fuel operating levels. The conversion also enabled ArcelorMittal to achieve a 25 percent increase in material throughput and a 50 percent reduction in scale formation (R&D Magazine, 2010).

Insulation of Furnaces

Replacing conventional insulating materials with ceramic low-thermal–mass insulation materials can reduce the heat losses through furnace walls. The potential energy savings for insulating a continuous furnace were estimated to range from 2 to 5 percent, or approximately 0.14 MMBtu/ton (0.16 GJ/tonne) of product. Capital costs were estimated to be \$14.1/ton (\$15.6/tonne) of product. The payback time is estimated as 31 years.

Walking Beam Furnace

A walking beam furnace represents the state-of-the-art of efficient reheating furnaces. In a walking beam furnace, the stock is placed on stationary ridges and a revolving beam walks the product along through the furnace until the exit where the beam returns to the furnace entrance. WCI Steel has a walking beam furnace that also employed a state-of-the-art combustion control. The use of this furnace at WCI Steel resulted in a reduction in electricity usage by 25 percent per ton produced and a reduction in overall fuel consumption by 37.5 percent per ton produced compared to three pusher-type furnaces.

Controlling Oxygen Levels and Variable-Speed Drives on Combustion Air Fans

Controlling oxygen levels and using VSDs on the combustion air fans on the reheating furnace helps to optimize combustion in the furnace. Excess air can substantially decrease combustion efficiency as it leads to excessive waste gases. Fuel-air ratios of the burners should therefore be checked regularly. The use of VSDs on combustion air fans on the reheating furnace also helps to control the oxygen level, especially as the load of the furnace may vary over time. The savings depend on the load factor of the furnace and the control strategies applied. Implementing a VSD on a combustion fan of a walking beam furnace at

Cardiff Rod Mill (UK) reduced the fuel consumption by 48 percent and had a payback period of 16 months. Energy savings can vary widely depending on the specific installation, but one conservative estimate place the savings at 10 percent, or approximately 0.28 MMBtu/ton (0.33 GJ/tonne) of product. The estimated investment costs were $0.72/ton ($0.79/tonne) of product. The payback time is estimated as 0.8 years.

Heat Recovery to the Product
In cases that it is not possible to hot-charge the slabs directly from the caster, energy can be recovered bringing exhaust gases that leave the high temperature portion of the process into contact with the relatively cool slabs. This will preheat the slab charge. In a plant-wide assessment of North Star Steel (Iowa) it was estimated that using furnace flue gases to preheat the charge to a moderate temperature of 840-1,020°F (450-550°C) would result in costs savings of about 32 percent. Another study reports a 50 percent reduction of the unit energy consumption of a heating furnace when charging semi-finished products at a temperature above 1,200°F (650°C) and a 70-80 percent reduction at charging temperatures above 1,800°F (980°C).

Waste Heat Recovery from Cooling Water
Waste heat can be recovered from the hot strip mill cooling water to produce low-pressure steam. Estimated fuel savings are 0.034 MMBtu/ton (0.04 GJ/tonne) of product, with a required increase in electricity consumption of 0.15 kWh/ton (0.0006 GJ/tonne) of product. Investment costs were estimated to be $1.2/ton ($1.3/tonne) of product. Operating and maintenance costs may increase by $0.10/ton ($0.11/tonne) of product. The payback time is estimated as over 50 years.

8. Cold Rolling and Finishing

Heat Recovery on the Annealing Line
Heat recovery can be accomplished by generating steam from recovered waste heat or by installing recuperative or regenerative burners in the annealing furnace. By instituting several measures to recover heat, including regenerative burners, insulation improvement, process management, and VSDs, energy consumption can be reduced by as much as 40 percent. This equates to an energy savings of approximately 0.26 MMBtu/ton (0.3 GJ/tonne) and 2.7 kWh/ton (0.011 GJ/tonne). Investment costs were estimated to be $3.8/ton ($4.2/tonne) based on one mill in The Netherlands. The payback time is estimated as 4 years.

Reduced Steam Use in the Pickling Line
Lids and/or floating balls can be added to the heated hydrochloric acid bath in the pickling line to reduce evaporation losses. Energy savings of up to 17 percent, or approximately 0.16 MMBtu/ton (0.19 GJ/tonne), have been estimated. Estimated capital costs were $4.0/ton ($4.4/tonne) of product. The payback time is estimated as 7 years.

Automated Monitoring and Targeting System
Power demands on the cold strip mill can be reduced by installing an automated monitoring and targeting system to improve operating efficiency. A system installed at one British steel mill reduced the energy demand of the cold rolling mill by approximately 15 to

20 percent, or approximately 54 kWh/ton (0.22 GJ/tonne). Installation costs were estimated to be $1.56/ton ($1.72/tonne) of product, or $0.92/ton ($1.0/tonne) of crude steel. The payback time is estimated as 0.8 years.

Inter-Electrode Insulation in Electrolytic Pickling Line

The existing industrial electrolytic steel pickling process is only 30 percent current efficient. This efficiency can be increased by reducing inter-electrode short circuit current with inter-electrode isolation. Experiments have shown that the current efficiency of the process is improved from 20 percent without insulation to 100 percent with it. Complete insulation does, however, lead to sludge accumulation in the compartments where the steel band is anodic, resulting in an inhomogeneous electrolyte and higher maintenance requirements. Use of an insulation which covers less than 66 percent of the electrolyte cross section area between the anode and cathode electrode groups offers a compromise as it results in a significant improvement in the process efficiency while maintaining good circulation and homogeneity of the electrolyte solution. The method is relatively easily applicable as a retrofit. No cost information was available in the original reference.

Continuous Annealing

A continuous annealing furnace makes it possible to integrate the conventional batch annealing process (i.e., electrolytic cleaning - annealing - cooling - temper rolling - recoiling) into one line. The use of such a furnace can lead to significant energy saving and productivity. For instance, for a particular continuous annealing furnace, the annealing time for one roll is approximately 30 minutes, as compared to approximately 10 days for the conventional bath process. In addition, fuel consumption is reduced by about 33 percent.

Considerable differences in fuel consumption exist between different types of cooling equipment used in continuous annealing: the suction cooling roll uses only 14 percent of the power used by a gas jet system. The installation of continuous annealing equipment demands relatively high investment costs. For example, a new (to be constructed) continuous annealing facility with a capacity of about 500,000 ton/year (450,000 tonne/year) in the Midwest has an estimated cost of $225 million. No information was available for payback time.

9. General Measures for Energy Efficiency Improvements

Preventive Maintenance

Training programs and good housekeeping programs help to decrease energy consumption throughout the plant. Some estimates place the energy savings at 2 percent of total energy use, or a fuel savings of approximately 0.39 MMBtu/ton (0.45 GJ/tonne) of product and an electricity savings of approximately 0.034 MMBtu/ton (0.04 GJ/tonne) of product. One estimate of annual operating costs was $16,600 per plant, or approximately $0.018/ton ($0.02/tonne) of crude steel.

Energy Monitoring and Management System

Energy monitoring and management systems help provide for optimal energy recovery and distribution between processes at the plant. These systems may reduce energy consumption by 0.5 percent, or fuel savings of approximately 0.10 MMBtu/ton (0.12 GJ/tonne) of product and electricity savings of approximately 0.0086 MMBtu/ton (0.01

GJ/tonne) of product. Based on a system installed at one plant in The Netherlands, the cost of a monitoring and management system was approximately $0.21/ton ($0.23/tonne) of crude steel based on an investment cost of $1.2 million. The payback time is estimated as 0.5 years.

Combined Heat and Power/Cogeneration

All steel plants require both electricity and steam to operate, which make them good candidates for combined heat and power (CHP), also known as cogeneration. Modern CHP systems can be based on gas turbines with a waste heat recovery boiler, combined cycles that integrate a gas turbine with a steam turbine for larger systems, or high pressure steam boilers (both fuel-fired or waste heat boilers) coupled with a steam turbine generator. The type and size of CHP system utilized depends on a variety of site-specific factors including the amount and quality of off-gases from the coke oven, blast furnace, and BOF; the steam requirements of the facility, and the economics of generating power on-site versus purchasing power from the grid. CHP capital costs can range from $900 to $2,500/kW depending on size and technology (EPA, 2007b). Estimates range from $20.6/ton ($22.7/tonne) of crude steel. The payback time is estimated as 6 years. Over thirty steel and coke plants have currently installed CHP systems (ICF, 2010). The newest coke plants all recover the heat from the battery stack to produce steam and/or electricity. Most integrated iron and steel plants use excess process fuel gases (BFG and COG) for CHP units. Opportunities for waste heat to power CHP exist with heat recovery in the sinter plant and in dry coke quenching.

High-Efficiency Motors

Due to the high number of motors at an iron and steel plant, a systems approach to energy efficiency should be considered. Such an approach should look for energy efficiency opportunities for all motor systems (e.g., motors, drives, pumps, fans, compressors, controls). An evaluation of energy supply and energy demand should be performed to optimize overall performance. A systems approach includes a motor management plan that considers at least the following factors:

- Strategic motor selection;
- Maintenance;
- Proper size;
- Adjustable speed drives;
- Power factor correction; and
- Minimize voltage unbalances.

One estimate of overall energy consumption by motors in the steel industry was 22 billion kWh. DOE has estimated that 12 percent of this energy could be saved through the use of more efficient equipment. One estimate places the potential energy savings from motor efficiency improvements at 0.3 MMBtu/ton (0.35 GJ/tonne) (Stubbles, 2000). Payback time is estimated as 1 to 3 years.

Motor management plans and other efficiency improvements can be implemented at existing facilities and should be considered in the design of new construction.

B. Energy Efficiency Options for Electric Arc Furnace Steelmaking

Opportunities to improve energy efficiency at EAF steelmaking facilities are described below. In addition, the casting, hot rolling, and other finishing processes at EAF facilities are similar to those described for integrated iron and steel plants. Consequently, the opportunities described earlier for these processes also apply to EAF facilities and are included in Table 2.

Improved Process Control (Neural Networks)

Process control can optimize operations and thereby significantly reduce electricity consumption as is demonstrated by many examples worldwide. Modern controls which use a multitude of sensors can help to achieve this to a greater extent than older controls. Control and monitoring systems for EAF are moving towards integration of real-time monitoring of process variables, such as steel bath temperature, carbon levels, and distance to scrap, along with real-time control systems for graphite injection and lance oxygen practice. As an example, neural networks systems analyze data and emulate the best controller and can thus help to reduce electricity consumption beyond that achieved through classical control systems. Neural networks can help achieve additional reductions in energy consumption over classic control systems. For EAFs, average power savings were estimated to be 8 percent, or 34.5 kWh/ton (0.14 GJ/tonne). Additionally, productivity increased by 9 to 12 percent, and electrode consumption was reduced by 25 percent. Capital costs were estimated to be $372,500 per furnace, with annual cost savings of approximately $1.4/ton ($1.5/tonne). The payback time is estimated as 0.5 years.

By monitoring the furnace exhaust gas flow rate and composition, the use of chemical energy in the furnace can be enhanced. Detailed investigation of the post-combustion of off-gases can be carried by an optical sensor. Using the monitored data as input for a control system, post-combustion of off-gases can be controlled online. Benefits of this practice include reduced electricity consumption, shorter power-on times, increased productivity, a decrease in production costs, a reduction of electrode consumption, reduced natural gas, oxygen and carbon consumption, and a reduction of refractory wear. It has been demonstrated that, if oxygen injected for post-combustion is continuously controlled by real-time data acquisition of CO and CO_2 concentrations in off-gases, a 50 percent increase in recovery rate of chemical energy in fumes can be achieved compared to operation based on predefined set-points.

A specific system that continuously measures CO, CO_2, H_2, and O_2 to control post-combustion was installed at the Hylsa's Planta Norte plant near Monterrey (Mexico) and by Nucor, Seattle (WA). The system led to reductions of 2 percent and 4 percent in electricity consumption, 8 percent and 16 percent in natural gas consumption, 5 percent and 16 percent in oxygen use, 18 percent and 18 percent in carbon charged and injected. At the same time, yield improved (between 1 percent and 2 percent), and electrode consumption decreased (3.5 percent and 16 percent), while productivity increased by 8 percent.

Adjustable Speed Drives

As flue gas flow varies over time, adjustable speed drives offer opportunities to operate dust collection fans in a more energy efficient manner energy can. Flue gas adjustable speed drives have been installed in various countries (e.g., Germany, UK). The electricity savings are estimated to be 15 kWh/ton (0.06 GJ/tonne), with a payback period of 2 to 3 years.

Although dust collection rates were reduced by 2-3 percent, total energy usage decreased by 67 percent. Capital costs were estimated to be $1.8/ton ($2/tonne). The payback time is estimated as 2 to 3 years.

Transformer Efficiency—Ultra-High–Power Transformers

Ultra-high–power (UHP) transformers help to reduce energy loss and increase productivity. The UHP furnaces are those with a transformer capacity of more than 700 kilovolt amps (kVA)/tonne heat size. The UHP operation may lead to heat fluxes and increased refractory wear, making cooling of the furnace panels necessary. This results in heat losses that partially offset the power savings. Total energy savings were estimated to be 15 kWh/ton (0.061 GJ/tonne). Many EAF operators have installed new transformers and electric systems to increase the power of the furnaces, e.g., Co-Steel (Raritan, NJ), SMI (Sequin, TX), Bayou Steel (Laplace, LA), and Ugine Ardoise (France). Capital costs were estimated to be $3.9/ton ($4.3/tonne). The payback time is estimated as 5.2 years.

Bottom Stirring/Stirring Gas Injection

Bottom stirring is accomplished by injecting an inert gas into the bottom of the EAF to increase the heat transfer in the melt. In addition, increased interaction between slag and melt leads to an increased liquid metal yield of 0.5 percent. Furnaces with oxygen injection are sufficiently turbulent, reducing the need for inert gas stirring. The increased stirring can lead to electricity savings of 10 to 20 kWh/ton (0.04 to 0.08 GJ/tonne), with net annual production cost reduction of $0.72 to $1.4/ton ($0.8 to $1.6/tonne). Taking into account the increased liquid steel yield may increase the cost savings to $1.3 to $3.1/ton ($1.4 to $3.4/tonne). Power savings were estimated to be 18 kWh/ton (0.072 GJ/tonne). Capital costs for retrofitting existing furnaces were estimated to be $0.85/ton ($0.94/tonne) for increased refractory costs and installing tuyeres, and annual costs for inert gas purchase was estimated to be $1.8/ton ($2.0/tonne). Productivity increases were estimated to reduce costs by $5.0/ton ($5.5/tonne). The payback time is estimated as 0.2 years.

Foamy Slag Practice

Foamy slag covers the arc and melt surface to reduce radiation heat losses. Foamy slag can be obtained by injecting carbon (granular coal) and oxygen or by lancing of oxygen only. Slag foaming increases the electric power efficiency by at least 20 percent in spite of a higher arc voltage. The net energy savings (accounting for energy use for oxygen production) are estimated at 5-7 kWh/ton (0.02-0.028 GJ/tonne) steel. Foamy slag practice may also increase productivity through reduced tap-to-tap times. Investment costs are about $14.1/ton ($15.6/tonne) capacity. Productivity increases may be equivalent to a cost savings of approximately $2.6/ton ($2.9/tonne) steel. The payback time is estimated as 4.2 years.

Oxy-Fuel Burners

Oxy-fuel burners are used on approximately half the EAFs in the U.S. These burners increase the effective capacity of the furnace by increasing the speed of the melt and reducing the consumption of electricity and electrode material, which reduces GHG emissions. The use of oxy-fuels burners has several beneficial effects: it increases heat transfer, reduces heat losses, reduces electrode consumption and, and reduces tap-to-tap time. Moreover, the injection of oxygen helps to remove different elements from the steel bath, like phosphorus, silicon and carbon. Steelmakers are now making wide use of stationary wall-mounted oxygen-gas burners and combination lance-burners, which operate in a burner mode during the initial part of the melting period. When a liquid bath is formed, the burners change over to a mode in which they act as oxygen lances. Electricity savings may range from 88-155 kWh/ft^3 (11-20 GJ/m^3) oxygen injected. Natural gas injection is typically 10 standard cubic feet per kilowatt hour, with energy savings ranging from 18-36 kWh/ton (0.72-0.14 GJ/tonne). Investment cost for modifying a 121 ton (110 tonne) EAF were estimated to be $6.8/ton ($7.5/tonne). Annual cost savings may be approximately $6.4/ton ($7.1/tonne) due to reduced tap-to-tap times. The payback time is estimated as 0.9 years.

Post-Combustion of the Flue Gases

Post-combustion is a process for utilizing the chemical energy in the CO and hydrogen evolving from the steel bath to heat the steel in the EAF ladle or to preheat scrap to 570-1,470°F (300-800°C). It reduces electrical energy requirements and increases the productivity of the EAF. Other benefits include reduction of baghouse emissions, reduction of the temperature of the off-gas system and minimization of high temperature spikes associated with rapid CO evolution. Post-combustion helps to optimize the benefits of oxygen and fuel injection. EAF operations that involve large amounts of charged carbon or pig iron are particularly suitable for implementation of CO post-combustion technology during scrap melting.

It is critical that post-combustion is done early at melt down while the scrap is still capable of absorbing the evolved heat. The injectors should be placed low enough to increase CO retention time in the scrap in order to transfer its heat. The oxygen flow should have a low velocity to promote mixing with the furnace gases and avoid both scrap oxidation and oxygen rebound from the scrap to the water cooled panels. The injectors should also be cooled extremely well as the post combustion area often gets overheated. In order to distribute the chemical energy uniformly and to make its utilization efficient, it is preferable to bifurcate the post combustion oxygen flow and to space out the injectors in the colder areas of the shell.

For a particular post-combustion system, electricity savings ranged from 6 to 11 percent and reductions in tap-to-tap time from 3 to 11 percent, depending on the operating conditions. No information was available for costs or payback time.

Direct Current Arc Furnace

The direct current (DC) arc furnace was pioneered in Europe, and these single-electrode furnaces have recently been commercialized in North America. The DC arc furnaces use DC rather than alternating current (AC). In a DC furnace one single electrode is used, and the bottom of the vessel serves as the anode. Based on the distinctive feature of using the heat and magnetic force generated by the current in melting, this arc furnace achieves an energy saving of approximately 5 percent in terms of power unit consumption in comparison with the 3-phase AC arc furnace. In addition, it also has other features, including higher melting efficiency and extended hearth life. Power consumption is 454-544 kWh/ton (1.8-2.2 GJ/tonne) molten steel. Electrode consumption is about half that with conventional furnaces. This corresponds to 2.4-4.9 lb/ton (1-2 kg/tonne) molten steel. This measure is applied to large furnaces only. Net energy savings were estimated to be 82 kWh/ton (0.32 GJ/tonne). However, compared to new AC furnaces, the savings are limited to 9-18 kWh/ton (0.036-

0.072 GJ/tonne). The additional investment costs over that of an AC furnace are approximately $5.5/ton ($6.1/tonne) capacity. The payback time is estimated as 0.7 years.

The design of the DC arc furnace also reduces noise and electrical flicker, increases efficiency, and reduces electrode consumption. As of 2007, there are eight DC powered EAFs operating in the U.S. and one in Mexico; most of these EAFs have been installed in the past 2 years, the oldest was installed in 1991. The manufacturers involved are Fuchs, NKK/United, MAN Ghh, and Voest-Alpine. Facilities that are currently using this new technology include Charter Steel, Florida Steel, Gallatin Steel, North Star Steel, and many Nucor plants (e.g., Blytheville, AR; Berkeley, SC; Decatur, AL; Hertford, NC; Norfolk, NE; Darlington, SC).

Scrap Preheating

Scrap preheating is performed either in the scrap charging baskets, in a charging shaft (shaft furnace) added to the EAF, or in a specially designed scrap conveying system allowing continuous charging during the melting process. Scrap preheating is used extensively in Japan, and the use of hot furnace gases for scrap preheating is now being applied in the U.S. Scrap preheating can save 4–50 kWh/ton (0.016-0.20 GJ/tonne) and reduce tap-to-tap times by 8 to 10 minutes. A prominent example of its application to new EAFs with continuous charging is the Consteel process, which is being used at Gerdau-Ameristeel plants in Charlotte, NC, Knoxville, TN, and Sayreville, NJ; and at Nucor plants in Darlington, SC, and Hertford, NC.

Preheating scrap reduces the power consumption of the EAF by using the waste heat of the EAF as the energy source for the preheat operation. The Consteel process consists of a conveyor belt that transports the scrap through a tunnel to the EAF. In addition to energy savings, the Consteel process can increase productivity by 33 percent, decrease electrode consumption by 40 percent, and reduce dust emissions. Electricity savings can be 54 kWh/ton (0.22 GJ/tonne), and investment costs were estimated to be $3.2 million for a capacity of 550,000 ton/yr (500,000 tonne/yr) or $7.1/ton ($7.8/tonne) of product. Annual costs savings were estimated to be $2.7/ton ($3.0/tonne). The payback time is estimated as 1.3 years.

Scrap Preheating, Post Combustion—Shaft Furnace (Fuchs)

Shaft-furnace technology (both single- and double-shaft furnaces) was pioneered by Fuchs in the late 1980s. Since 2005, the VAI Fuchs furnace has been known as SIMETALCIS EAF. With the single shaft furnace, up to 70 kWh/ton (0.28 GJ/tonne) liquid steel of electric power can be saved. The finger shaft furnace[9] allows energy savings up to 100 kWh/ton (0.40 GJ/tonne) liquid steel, which is about 25 percent of the overall electricity input into the furnace. The exact energy savings depend on the scrap used, and the degree of post-combustion (oxygen levels). For the finger shaft furnace tap-to-tap times of about 35 minutes are achieved, which is about 10-15 minutes less compared to EAF without efficient scrap preheating. The process may reduce electrode consumption, improve yield by 0.25 to 2 percent, increase productivity by 20 percent, and decrease flue gas dust emissions by 25 percent. Retrofit costs were estimated to be $8.5/ton ($9.4/tonne) for an existing 110-ton (100-tonne) furnace. Production cost savings may amount to $6.1/ton ($6.7/tonne). The payback time is estimated as 1 year.

---

[9] The most efficient shaft-furnace design is the finger-shaft furnace, which employs a unique scrap retaining system with fingers to preheat 100 percent of the scrap charge using the hot flue gases.

Engineered Refractories

Refractories in EAF have to withstand extreme conditions such as temperatures over 2,900°F (1,600°C), oxidation, thermal shock, erosion and corrosion. These extreme conditions generally lead to an undesired wear of refractories. Refractories can be provided by a controlled microstructure: alumina particles and mullite microballoons coated uniformly with carbon and carbides. The refractories can be either sintered or cast and can therefore be used in a wide range of components at EAF mills (e.g., furnace, ladle furnace, vessels). The refractories can reduce ladle leakages and the formation of slag in transfer operations with savings of 10 kWh/ton (0.04 GJ/tonne) steel.

Airtight Operation

A large amount of air enters the EAF: around 1,000,000 ft^3 (30,000 m^3) in a standard EAF of 165 tons (150 tonnes) of steel with a heat duration of 1 hour. This air is at ambient temperature, and the air's nitrogen and non-reactive oxygen are heated in the furnace and exit with the fumes at high temperature (around 1,800°F) (980°C), resulting in significant thermal losses. Based on the results of pilot scale trials with a 7 ton (6 tonne) EAF at Arcelor Research, the potential benefit for an industrial furnace with an airtight process including a post-combustion practice and an efficient fume exhaust control are about 100 kWh/ton (0.4 GJ/tonne) for an industrial furnace having a current electric consumption of 450 kWh/ton (1.8 GJ/tonne). About 80 percent of the savings can be attributed to a reduction of energy losses in the fumes. The remaining 20 percent are accounted for by reduced thermal losses due to a reduced tap-totap time. The exhaust gas can be used as a fuel in the post-combustion chamber and reduces the amount of natural gas needed for the burner.

Contiarc® Furnace

The *Contiarc* furnace is fed continuously with material in a ring between the central shaft and the outer furnace vessel, where the charged material is continuously preheated by the rising process gas in a counter-current flow, while the material continuously moves down. Located below the central shaft is a "free-melting volume" in the form of a cavern. Advantages of the Contiarc furnace include (1) reduced energy losses (200 kWh/ton or 0.8 GJ/tonne less than with conventional furnace systems), (2) waste gas and dust volumes are considerably reduced, which results in a lower capacity for the gas cleaning system and also lower electric power consumption (23 kWh/ton or 0.091 GJ/tonne), (3) gas-tight furnace enclosure captures all primary and nearly all secondary emissions, and (4) reduced electrode consumption (about 1.8 lb/ton or 0.9 kg/tonne less than a typical AC furnace).

Flue Gas Monitoring and Control

The use of VSDs can reduce energy usage of the flue gas fans, which in turn reduces the losses in the flue gas. Electricity savings were estimated to be 13.6 kWh/ton (0.054 GJ/tonne) with a payback period of 2 to 4 years. Capital costs were estimated to be $2.8/ton ($3.1/tonne).

Eccentric Bottom Tapping

Eccentric bottom tapping leads to slag-free tapping, shorter tap-to-tap times, reduced refractory and electrode consumption, and improved ladle life. Energy savings were estimated to be 13.6 kWh/ton (0.054 GJ/tonne). Modification costs for a Canadian plant were $3.3

million for a furnace with an annual production capacity of 760,000 tons (690,000 tonnes) or $4.5/ton ($5.0/tonne). The payback time is estimated as 7 years.

Twin-Shell Furnace

A twin-shell furnace includes two EAF vessels with a common arc and power supply system. The system increases productivity by decreasing tap-to-tap time, and reduces energy consumption by reducing heat losses. A twin-shell furnace may save 17 kWh/ton (0.068 GJ/tonne) compared to a single-shell furnace. Production costs are expected to be $1.8/ton ($2.0/tonne) lower that a single-shell furnace, and the investment costs are expected to be approximately $8.5/ton ($9.4/tonne) over that of a single-shell furnace. The payback time is estimated as 3.5 years.

C. Long-Term Opportunities to Reduce CO_2 Emissions (Worrell et al., 2009)

Increasing the primary energy efficiency of the production process will lead to a reduction in specific CO_2 emission. The energy efficiency measures that were discussed in the previous chapters therefore offer opportunities to reduce emissions. These reductions can be determined by multiplying the reductions in fuel and electricity usage with emission factors. Note that emission factors are dependent on the type of fuel used and the way electricity is generated.

A way to decrease CO_2 emissions from the blast furnaces ironmaking process is using hydrogen bearing materials such as steam, natural gas and waste plastics to substitute coke and coal. Apart from CO_2 emission reductions due to energy efficiency of resource utilization, there are several emerging technologies to mitigate emissions. However, none are currently commercially available or used at commercial scale. Development of these technologies may take decades.

The global steel industry collaborates in the ULCOS project to find opportunities to dramatically reduce CO_2 emissions from iron and steelmaking. ULCOS is a consortium of 48 European companies and organizations from 15 European countries that have launched a cooperative R&D initiative to enable drastic reduction in CO_2 emissions from steel production. The aim of the ULCOS program is to reduce the CO_2 emissions by at least 50 percent.

ULCOS has selected four process concepts that could lead to a reduction of CO_2 emissions by more than half compared to current best practice. The following are the four breakthrough technologies identified:

- Top gas recycling blast furnace with carbon capture and storage (CCS);
- HIsarna with CCS;
- Advanced direct reduction with CCS; and
- Electrolysis.

Traditionally, carbon from fossil fuels is used in the steel industry to provide the chemical function of reducing oxide ores.

This function could also be performed hydrogen or carbon-free electricity, or wood:

- Hydrogen reduction of iron ore has steam as a gas product instead of CO_2. Hydrogen can be produced from natural gas, by electrolysis of seawater, etc. Limitations are not technical, as technologies in the area of pre-reduction are very mature, and are related to the volatile issue of resource depletion in the longer term. Research projects are underway in different countries. In the U.S. the AISI, U.S. DOE, and the University of Utah investigate a process called Hydrogen Flash Smelting.
- The concept of using wood to make iron in a charcoal blast furnace is presently being applied in Brazil.
- Electrolysis, which leads directly to final products, is to be compared to a whole conventional mill, which has an energy consumption of 15 to 20 GJ/t liquid steel, with a similar order of magnitude. The technology might be attractive in terms of CO_2 emissions, and if the carbon content of electricity is sufficiently low. The most promising options for electrolysis are aqueous alkaline electrolysis, also called electrowinning, and iron ore pyroelectrolysis. Both technologies have already been shown possible at very small scale while commercial application may still be decades away. In the U.S. the Massachusetts Institute of Technology (MIT), AISI, and U.S. DOE jointly investigate the opportunities of electrolysis processes for ironmaking.

HIsarna is a technology based on bath-smelting. It combines coal preheating and partial pyrolysis in a reactor, a melting cyclone for ore melting and a smelter vessel for final ore reduction and iron production. It requires significantly less coal usage and thus reduces the amount of CO_2 emissions. Furthermore, it is a flexible process that allows partial substitution of coal by biomass, natural gas or even hydrogen. The HIsarna process is based on the Cyclone Converter Furnace developed by Hoogovens (The Netherlands). The Cyclone Converter Furnace technology incorporates the results of earlier AISI projects to develop convertor-based reduction processes. A pilot plant will be operational in early 2010. Additional work is continuing on using CCS and biomass technology in combination with HIsarna.

Carbon Capture and Storage

Carbon capture and storage involves separation and capture of CO_2 from the flue gas, pressurization of the captured CO_2, transportation of the CO_2 via pipeline, and finally injection and long-term geologic storage of the captured CO_2. Several different technologies, at varying stages of development, have the potential to separate and capture CO_2. Some have been demonstrated at the slip-stream or pilot-scale, while many others are still at the bench-top or laboratory stage of development.

In 2010, an Interagency Task Force on Carbon Capture and Storage was established to develop a comprehensive and coordinated Federal strategy to speed the commercial development and deployment of clean coal technologies. The Task Force was specifically charged with proposing a plan to overcome the barriers to the widespread, cost-effective deployment of CCS within 10 years, with a goal of bringing 5 to 10 commercial demonstration projects online by 2016.

As part of its work, the Task Force prepared a report that summarizes the state of CCS and identified technical and non-technical barriers to implementation. The development status of CCS technologies is thoroughly discussed in the Task Force report. For additional information on the Task Force and its findings on CCS as a CO_2 control technology, go to: http://www.epa.gov/climatechange/policy.

D. Emerging Technologies

According to AISI, the greatest potential for reducing the energy intensity of steelmaking lies with development of new transformational technologies and processes. Examples of such transformational R&D efforts (applicable both to integrated and EAF steelmaking) include the following: (1) molten oxide electrolysis (under development at MIT); (2) ironmaking by flash smelting using hydrogen (under development at the University of Utah); and (3) the paired straight hearth (PSH) furnace (under development at McMaster University in Ontario, Canada).

AISI lists the following additional areas as important R&D opportunities for EAF steelmaking: improved processes for low-grade scrap recovery, and sensible heat recovery from slag, fumes, and off-gases (EPA, 2007a).

The R&D opportunities noted in the DOE study (EPA, 2007a) include increasing the efficiency of melting processes (0.4 MMBtu/ton or 0.47 GJ/tonne), integration of refining functions and reductions of heat losses prior to casting (0.35 MMBtu/ton or 0.41 GJ/tonne), economical heat capture from EAF waste gas (0.26 MMBtu/ton or 0.30 GJ/tonne), purification and upgrading to scrap, and effective use of slag and dust. Casting and rolling opportunities (applicable both to integrated and EAF steelmaking) include the reduction of heat losses from cast products prior to rolling and/or reheating (0.75 MMBtu/ton or 0.87 GJ/tonne) and thin-strip casting (0.5–0.7 MMBtu/ton [0.3-0.8 GJ/tonne]) (EPA, 2007a). Appendix B provides more details on active and completed research projects conducted by DOE and DOE's partnerships.

1. Integrated DRI/EAF Steelmaking

The Essar Group, which acquired Minnesota Steel in late 2007, was constructing a $1.6-billion steel-making facility on the Mesabi iron ore range in Minnesota that would be the first of its type (from iron ore to steel product at the mine site) (Essar Steel Minnesota LLC, 2010). However, construction has been halted due to economic reasons. This new plant will produce 4.1 million tons/yr (3.7 million tonnes/yr) of direct reduced iron (DRI) pellets, most of which will be processed in EAFs to produce 2.8 million tons/yr (2.5 million tones/yr) of steel slabs.

This integrated steel-making route requires less energy and produces lower emissions than traditional integrated steelmaking (i.e., coke battery, blast furnace, BOF). A DOE (2008) report claims the following reduction in emissions relative to traditional steelmaking:

Pollutant	Percent Reduction
CO	96
Volatile organic compounds	87
Sulfur dioxide	78
NO_x	65
Mercury	58
CO2	41

2. EAF Steelmaking at an Integrated Plant

Wheeling-Pittsburgh Steel (now owned by Severstal) installed a state-of-the art EAF in December 2004 to replace its BOF for steelmaking. This is the first application of an EAF at an integrated mill to convert molten iron from the blast furnace into steel. The EAF is continuously charged with molten iron and scrap (BOFs and most EAFs are batch processes), can use up to 100 percent scrap, recovers heat from the EAF exhaust to preheat the scrap, and produces 330 tons/hr (300 tonnes/hr) of steel. After the EAF was installed, one of the two blast furnaces was shut down. The company claimed there were significant cost and environmental benefits from the conversion (Tenova, 2010).

APPENDIX A. ADDITIONAL BACKGROUND DETAILS FOR THE IRON AND STEEL INDUSTRY

A.1. Plants and Locations

Tables A-1, A-2, and A-3 list the operating plants (in 2009) for the 19 coke plants, 18 integrated plants, and 95 EAF plants, respectively. Several integrated plants are clustered around the Great Lakes, which facilitates the delivery of taconite (processed iron ore) by waterway from iron ore mines on the Mesabi iron ore range in Minnesota and Michigan. Most coke plants are located at or near the integrated iron and steel plants. Several integrated plants are in non-attainment areas for particulate matter with a diameter less than 2.5 micrometers (PM2.5).

Table A-1. List of Coke Plants (AIST, 2009; EPA, 2001a)

No.	Company Name	City	State	No. of Batteries	Coke Capacity (tpy)	Type of Plant	Type of Coke	Type of Battery	Cogeneration
1	ABC Coke	Tarrant	AL	3	699,967	Merchant	Foundry	By-product	No
2	AK Steel[a]	Middletown	OH	1	1,200,000	Captive	Furnace	By-product	No
3	AK Steel[a]	Ashland	KY	2	2,000,000	Captive	Furnace	By-product	No
4	ArcelorMittal[a]	Burns Harbor	IN	2	1,877,000	Captive	Furnace	By-product	No
5	EES Coke/DTE Energy	Ecorse	MI	1	1,000,000	Merchant	Furnace	By-product	No
6	Erie Coke	Erie	PA	2	214,951	Merchant	Foundry	By-product	No
7	Gateway Energy & Coke	Granite City	IL	3	650,000	Merchant	Furnace	Nonrecovery	Yes[b]
8	Koppers	Monessen	PA	2	372,581	Merchant	Furnace	By-product	No
9	Mittal Steel	Warren*	OH	1	519,000	Captive	Furnace	By-product	No
10	Severstal[a]	Mingo Junction	OH	4	1,346,000	Captive	Furnace	By-product	No
11	Shenango	Neville Island	PA	1	514,779	Merchant	Furnace	By-product	No
12	Sloss Industries	Birmingham	AL	3	451,948	Merchant	Foundry	By-product	No
13	Haverhill Coke	Haverhill	OH	4	1,100,000	Merchant	Furnace	Nonrecovery	Yes[b]
14	Indiana Harbor Coke	East Chicago	IN	4	1,300,000	Merchant	Furnace	Nonrecovery	Yes[b]
15	Jewell Coke and Coal	Vansant	VA	6	745,000	Merchant	Furnace	Nonrecovery	No
16	Tonawanda	Tonawanda	NY	1	268,964	Merchant	Foundry	By-product	No
17	U.S. Steel	Clairton	PA	12	5,573,185	Captive	Furnace	By-product	Yes[c]
18	U.S. Steel[a]	Granite City	IL	2	584,000	Captive	Furnace	By-product	No
19	U.S. Steel[a]	Gary	IN	4	2,249,860	Captive	Furnace	By-product	No
	Total			58	22,700,000				

Note: tpy = tons per year.
[a] Located at an integrated iron and steel plant (see Table A-2).
[b] Recovery of waste heat from the battery stack.
[c] Combustion of excess COG.

Table A-2. List of Integrated Iron and Steel Plants (AIST, 2009; EPA, 2001b)

No.	Company	City	State	No. of BOFs	Steelmaking Capacity (tpy)	No. of Blast Furnaces	Iron Capacity (tpy)	No. of Coke Batteries	Coke Capacity (tpy)	Sinter Capacity (tpy)
1	AK Steel	Ashland	KY	2	2,100,000	1	2,000,000	2	2,000,000	
2	AK Steel	Middletown	OH	2	2,640,000	1	2,300,000	1	1,200,000	
3	ArcelorMittal	Burns Harbor	IN	3	5,600,000	2	5,460,000	2	1,877,000	2,800,000
4	ArcelorMittal	Cleveland	OH	4	5,100,000	2	3,100,000			
5	ArcelorMittal	East Chicago	IN	2	3,800,000	2	3,100,000			1,200,000
6	ArcelorMittal	East Chicago	IN	4	6,250,000	3	6,500,000			1,100,000
7	ArcelorMittal	Weirton	WV	2	3,000,000	2	2,700,000			
8	ArcelorMittal	Riverdale	IL	2	1,100,000					
9	Republic Engineered Products	Lorain	OH	2	2,700,000	1	1,460,000			
10	Severstal	Dearborn	MI	2	4,100,000	1	2,190,000			
11	Severstal	Mingo Junction	OH	2	2,400,000	1	1,350,000	4	1,346,000	
12	Severstal	Sparrows Point	MD	2	3,375,000	1	3,100,000			3,430,000
13	Severstal	Warren	OH	2	2,040,000	1	1,460,000			
14	U.S. Steel	Braddock	PA	2	2,957,000	2	2,300,000			
15	U.S. Steel	Ecorse	MI	2	3,900,000	3	4,150,000			
16	U.S. Steel	Fairfield	AL	3	2,920,000	1	2,190,000			
17	U.S. Steel	Gary	IN	6	8,730,000	4	7,340,000	4	2,249,860	4,400,000
18	U.S. Steel	Granite City	IL	2	3,000,000	2	2,400,000	2	584,000	
	Totals			46	65,712,000	30	53,100,000	15	9,256,860	12,930,000

Note: tpy = tons per year.

Table A-3. List of EAF Steel Plants (AIST, 2009)

No.	Company Name	City	State	Steel Type	No. of EAFs	Capacity (tpy)
1	AK Steel	Butler	PA	C, S	3	1,000,000
2	AK Steel	Mansfield	OH	C, S	2	700,000
3	Allegheny Technologies	Brackenridge	PA	S	2	550,000
4	Allegheny Technologies	Midland	PA	S	2	600,000
5	Alton Steel	Alton	IL	C	1	700,000
6	Arcelor Mittal	Coatesville	PA	C, H, S	1	880,000
7	Arcelor Mittal	Georgetown	SC	C	1	550,000
8	Arkansas Steel	Newport	AR	C	1	130,000
9	Bayou Steel Corp.	LaPlace	LA	C	2	800,000
10	BetaSteel Corp.	Portage	IN	C	1	500,000
11	Border Steel Mills	Vinton	TX	C	2	250,000
12	Carpenter Technology	Reading	PA	H, S	6	180,000
13	Cascade Steel Rolling Mills	McMinnville	OR	C	1	750,000
14	Champion Steel Co.	Orwell	OH	C, H, S	1	6,000
15	Charter Steel	Cleveland	OH	C	1	700,000
16	Charter Steel	Saukville	WI	C	1	625,000
17	CMC Mills	Birmingham	AL	C, S	1	800,000
18	CMC Mills	Cayce-W.	SC	C	1	800,050
19	CMC Mills	Seguin	TX	C	1	1,000,000
20	Crucible Specialty Metals	Solvay	NY	H, S	1	50,000
21	Electralloy	Oil City	PA	H, S	1	60,000
22	Ellwood Quality Steels	New Castle	PA	C, H, S	1	400,000
23	Erie Forge and Steel Inc.	Erie	PA	C, H, S	3	385,000
24	Evraz Claymont Steel	Claymont	DE	C	1	490,000
25	Evraz Rocky Mountain Steel	Pueblo	CO	C	1	1,200,000
26	Finkl, A., & Sons	Chicago	IL	C	2	90,000
27	Geradu MacSteel	Baldwin	FL	C	1	1,100,000
28	Gerdau	Sand Springs	OK	C	2	600,000
29	Gerdau Ameristeel	Perth Amboy	NJ	C	1	800,000
30	Gerdau Ameristeel	Sayreville	NJ	C	1	750,000
31	Gerdau Ameristeel	Jackson	TN	C	1	892,000
32	Gerdau Ameristeel	Knoxville	TN	C	1	600,000
33	Gerdau Ameristeel	Vidor	TX	C	1	1,002,000
34	Gerdau Ameristeel	Monroe	MI	C	1	500,000
35	Gerdau Ameristeel (Gallatin Steel)	Warsaw	KY	C	2	1,600,000

Table A-3. (Continued)

No.	Company Name	City	State	Steel Type	No. of EAFs	Capacity (tpy)
36	Gerdau Macsteel	Fort Smith	AR	C	2	500,000
37	Gerdau Macsteel	Cartersville	GA	C, S	1	658,000
38	Gerdau Macsteel	Wilton	IA	C	1	917,000
39	Gerdau Macsteel	Jackson	MI	C, H	2	290,000
40	Gerdau Macsteel	St. Paul	MN	C	1	843,000
41	Gerdau Macsteel	Charlotte	NC	C	1	450,000
42	Haynes International	Kokomo	IN	H, S	2	20,000
43	Hoeganaes Corp.	Cinnaminson	NJ	C	1	112,000
44	Hoeganaes Corp.	Gallatin	TN	C	1	500,000
45	Keystone Steel & Wire Co	Peoria	IL	C	1	1,000,000
46	Kobelco Metal Powder of	Seymore	IN	C	1	63,000
47	LeTourneau Inc.	Longview	TX	C, H	2	125,000
48	Lone Star Steel	Lone Star	TX	C	2	265,000
49	Mittal (formerly Bethlehem Steel)	Steelton	PA	C	1	1,200,000
50	Mittal (formerly Ispat	East Chicago	IN	C	1	500,000
51	National Forge	Irvine	PA	C, H, S	1	58,000
52	North American Hoganas	Hollsopple	PA	H, S	1	300,000
53	North American Stainless	Ghent	KY	S	2	1,600,000
54	North Star BHP Steel, LLP	Delta	OH	C	1	1,800,000
55	NS Group Inc./Koppel Steel	Ambridge	PA	C	1	550,000
56	Nucor Corp.	Birmingham	AL	C	1	600,000
57	Nucor Corp.	Trinity	AL	C	2	2,400,000
58	Nucor Corp.	Tuscaloosa	AL	C	1	1,300,000
59	Nucor Corp.	Blytheville	AR	C	2	3,000,000
60	Nucor Corp.	Bourbonnais	IL	C	1	850,000
61	Nucor Corp.	Crawfordsvill	IN	C, H	2	2,400,000
62	Nucor Corp.	Flowood	MS	C	1	550,000
63	Nucor Corp.	Cofield	NC	C	1	1,400,000
64	Nucor Corp.	Norfolk	NE	C	1	1,100,000
65	Nucor Corp.	Auburn	NY	C	1	600,000
66	Nucor Corp.	Marion	OH	C	1	450,000
67	Nucor Corp.	Darlington	SC	C	1	1,050,000
68	Nucor Corp.	Huger	SC	C	2	3,450,000
69	Nucor Corp.	Jewett	TX	C	1	1,250,000
70	Nucor Corp.	Plymouth	UT	C	2	1,120,000

No.	Company Name	City	State	Steel Type	No. of EAFs	Capacity (tpy)
71	Nucor Corp.	Seattle	WA	C	1	1,100,000
72	Nucor-Yamato Steel Co.	Blytheville	AR	C, H, S	2	2,750,000
73	Oregon Steel Mills	Portland	OR	C,H	1	499,000
74	Republic Engineered Steels	Canton	OH	C	2	850,000
75	Severstal	Columbus	MS	C	1	1,700,000
76	Severstal	Steubenville	OH	C	1	2,400,000
77	SSAB North American Division	Axis	AL	C	1	1,250,000
78	SSAB North American	Muscatine	IA	C, H	1	1,250,000
79	Standard Steel	Latrobe	PA	C, H	1	59,000
80	Standard Steel	Burnham	PA	C, H, S	3	231,000
81	Steel Dynamics Inc.	Butler	IN	C	2	3,000,000
82	Steel Dynamics Inc.	Pittsboro	IN	C	1	720,000
83	Steel Dynamics Inc.	Columbia	IN	C, H	2	2,000,000
84	Steel Dynamics Inc.	Roanoke	VA	C	1	650,000
85	Steel of West Virginia, Inc.	Huntington	WV	C	2	265,000
86	Sterling Steel Co., LLC	Sterling	IL	C	1	1,100,000
87	TAMCO	Rancho	CA	C	1	750,000
88	Timken Co.	Latrobe	PA	C, H	2	60,000
89	Timken Co., Faircrest Plant	Canton	OH	C, H	1	870,000
90	Timken Co., Harrison Plant	Canton	OH	C, S	3	358,000
91	TXI Chaparral Steel	Midlothian	TX	C, H, S	2	2,000,000
92	TXI Chaparral Steel	Petersburg	VA	C, H, S	1	1,200,000
93	Union Electric Steel Corp.	Carnegie	PA	C, H	1	35,000
94	Universal Stainless & Alloy	Bridgeville	PA	H, S	1	150,000
95	V&M Star	Youngstown	OH	C, H	1	650,000
	Total					81,208,000

Note: C = carbon, H = high alloy, S = stainless, tpy = tons per year.

A.2. Greenhouse Gas Emission Estimates

Table A-4 provides preliminary estimates of GHG emissions from various iron and steel operations (based primarily on emissions factors and 2007 production rates). More precise and process-specific estimates of GHG emissions will be available in early 2011 when facilities report under EPA's mandatory reporting rule for GHGs. Blast furnaces and non-recovery coke plants are the largest emitters on a per process basis. Note that due to the economic downturn and lower production rates, GHG emissions may be lower for 2008-2010.

Table A-4. Preliminary Estimates of GHG Emissions (2007)

Type of Facility	Number of Facilities	Emissions (million tonnes of CO2/yr)				
		Process Units	Miscellaneous Combustion Units	Indirect Emissions (Electricity)	Facility Total	Average per Plant
Coke and EAF						
By-product coke (standalone)	9	2.8	2.7		5.5	0.6
Nonrecovery coke	3	3.0			3.0	1.0
EAF facilities	92	4.8	18.6	22.8	46.2	0.5
Integrated Iron and Steel Plants:						
By-product coke (co-located)	6	1.2			1.2	0.2
Blast furnace	17	23.9			23.9	1.4
BOF	18	4.4			4.4	0.2
Sinter plant	5	2.7			2.7	0.5
Totals						
All Integrated Iron and Steel	18	32.2	16.8	6.5	55.5	3.1
All Facilities	130	42.8	38.0	29.3	110.1	0.8

Note: yr = year.

A.3. Overview of the Iron and Steel Industry (EPA, 2001b; 2008b)

The iron and steel industry in the U.S. is the third largest in the world (after China and Japan), accounting for approximately 8 percent of the world's raw iron and steel production and supplying several industrial sectors, such as construction (building and bridge skeletons and supports), vehicle bodies, appliances, tools, and heavy equipment. The iron and steel industry actually includes three industries that have been traditionally treated as three different source categories: cokemaking, integrated iron- and steelmaking, and EAF steelmaking (secondary steelmaking that primarily recycles steel scrap).

A.3.1. Sinter Production (EPA, 2001b; 2008b)

Sintering is a process that recovers the raw material value of many waste materials generated at iron and steel plants that would otherwise be landfilled or stockpiled. An important function of the sinter plant is to return waste iron-bearing materials to the blast furnace to produce iron. Another function is to provide part or all of the flux material (e.g., limestone, dolomite) for the ironmaking process. There are currently five iron and steel plants with sintering operations, and all of the sinter plants are part of an integrated iron and steel plant.

Sintering is a continuous process. Feed material to the sintering process includes ore fines, coke, reverts (including blast furnace dust, mill scale, and other by-products of steelmaking), recycled hot and cold fines from the sintering process, and trim materials (e.g., calcite fines, and other supplemental materials needed to produce a sinter product with prescribed chemistry and tonnage). The materials are proportioned and mixed to prepare a chemically uniform feed to the sinter strand, so that the sinter will have the qualities desired for satisfactory operation of the blast furnace. The chemical quality of the sinter is often assessed in terms of its basicity, which is the percent total basic oxides divided by the percent total acid oxides {[CaO + MgO (calcium oxide plus magnesium oxide)])/[(SiO_2 + Al_2O_3) (silicon dioxide plus aluminum oxide)]}; sinter basicity is generally 1.0 to 3.0. The relative amounts of each material are determined based on the desired basicity, the rate of consumption of material at the sinter strand, the amount of sinter fines that must be recycled, and the total carbon content needed for proper ignition of the feed material.

The sintering machine accepts feed material and conveys it down the length of the moving strand. Near the feed end of the grate, the bed is ignited on the surface by gas burners and, as the mixture moves along on the traveling grate, air is pulled down through the mixture to burn the fuel by downdraft combustion; either COG or natural gas may be used for fuel to ignite the undersized coke or coal in the feed.

On the underside of the sinter strand is a series of windboxes that draw combusted air down through the material bed into a common duct, leading to a gas-cleaning device. The fused sinter is discharged at the end of the sinter strand, where it is crushed and screened. The sinter product is cooled in open air or in a circular cooler with water sprays or mechanical fans. The cooled sinter is crushed and screened a final time, and then is sent to be added or "charged" to the blast furnaces.

The primary emissions point of interest for the sinter plant is the stack that discharges the windbox exhaust gases after gas cleaning. The CO_2 is formed from the fuel combustion (COG or natural gas) and from carbon in the feed materials, including coke fines and other carbonaceous materials. Based on the Intergovernmental Panel on Climate Change (IPCC) emissions factor of 0.2 ton of CO_2/ton of sinter (0.2 tonne of CO_2/tonne of sinter) and the production of 14.7 million tons (13.3 million tonnes) of sinter in 2007, CO_2 emissions are estimated at 3.0 million tons (2.7 million tonnes) of CO_2/year. However, GHG emissions from sinter plants may vary widely over time as a consequence of variations in the fuel inputs and other feedstock, especially in the types and quantities of iron-bearing materials that are recycled. Because both natural gas and COG contain methane (CH_4), when these gases are burned, a small amount of the unburned CH_4 is emitted with the exhaust gases. Consequently, sinter plants (and any other process that burns fuels that contain CH_4) also emit a small amount of CH_4.

A.3.2. Blast Furnace Iron Production (EPA, 2001b; 2008b) Blast Furnace Operation

Iron is produced in blast furnaces by the reduction of iron-bearing materials with a hot gas. The large, refractory-lined furnace is charged through its top with iron ore pellets, sinter, flux (limestone and dolomite), and coke, which provides the fuel and forms a reducing atmosphere in the furnace. Many modern blast furnaces also inject pulverized coal or other sources of carbon to reduce the quantity of coke required. Iron oxides, coke, coal, and fluxes react with the heated blast air injected near the bottom of the furnace to form molten reduced iron, CO, and slag, which is a molten liquid solution of silicates and oxides that solidifies upon cooling. The molten iron and slag collect in the hearth at the base of the furnace. The by-product gas is collected at the top of the furnace and is recovered for use as fuel.

The production of 1 ton (0.91 tonne) of iron requires approximately 1.4 tons (1.3 tonnes) of ore or other iron-bearing material; 0.5 to 0.65 ton (0.45 to 0.59 tonne) of coke and coal; 0.25 ton (0.23 tonne) of limestone or dolomite; and 1.8 to 2 tons (1.6 to 1.8 tonnes) of air. By-products consist of 0.2 to 0.4 ton (0.18 to 0.36 tonne) of slag and 2.5 to 3.5 tons (2.3 to 3.2 tonnes) of BFG containing up to 0.05 tons (0.045 tonnes) of dust.

The molten iron and slag are removed (also called tapped), or cast, from the furnace in a semi-continuous process with 6 to 14 taps per day. The casting process begins by drilling a taphole into the clay-filled iron notch at the base of the hearth. During casting, molten iron flows into long troughs or "runners" that lead to transport containers, called "ladles." Slag also flows from the furnace and is directed through separate runners to a slag pit adjacent to the casthouse or into slag pots for transport to a remote slag pit. At the end of tapping, the taphole is replugged with clay. The area around the base of the furnace, including all iron and slag runners, is enclosed by a casthouse. The molten iron is transferred to a refractory-lined rail car (also called a "torpedo" car because of it shape) and is then sent to the BOF shop. The hot metal is then poured from the torpedo cars into the BOF shop ladle; which is referred to as "hot-metal transfer" or "reladling." Hot-metal transfer generally takes place under a hood to capture PM emissions, including kish (flakes of carbon), which is formed during the process.

Blast Furnace Gas

The BFG by-product, which is collected from the furnace top, has a low heating value and is composed of nitrogen (approximately 60 percent), CO (28 percent), and CO_2 (12 percent). Because of its high CO content, this BFG is used as a fuel within the steel plant. However, before BFG can be efficiently oxidized, the gas must be cleaned of dust or PM. Initially, the gases pass through a settling chamber or a dry cyclone to remove approximately 60 percent of the PM. Next, the gases undergo a one- or two-stage cleaning operation. The primary cleaner is normally a wet scrubber, which removes approximately 90 percent of the remaining PM. The secondary cleaner is a high-energy wet scrubber (usually a venturi) or an electrostatic precipitator, either of which can remove up to 90 percent of the PM that eludes the primary cleaner. Together, these control devices provide a clean gas of less than 0.02 grains per cubic foot (gr/ft^3) (0.05 grams per cubic meter [g/m^3]). A portion of this gas is fired in the blast furnace stoves that are used to preheat the air going into the blast furnace, and the remainder is used in other plant operations.

There are generally three to four stoves per blast furnace. Before the blast air is delivered to the blast furnace from the stoves, it is further preheated by passing it through a regenerator (heat exchanger) that uses some of the energy of the blast furnace off-gas that would otherwise have been lost. The additional thermal energy returned to the blast furnace (as heat) decreases the amount of fuel that has to be burned for each unit of hot metal and improves the efficiency of the process. In many furnaces, the off-gas is enriched by the addition of a fuel with much higher calorific value, such as natural gas or COG, to obtain even higher hot-blast temperatures. This decreases the fuel requirements and increases the hot-metal–production rate to a greater extent than is possible when burning BFG alone to heat the stoves.

Iron Preparation Hot-Metal Desulfurization

Sulfur in the molten iron is sometimes reduced before charging into the steelmaking furnace by adding reagents, such as soda ash, lime, and magnesium, in a process known as desulfurization. Injection of the reagents is accomplished pneumatically with either dry air or nitrogen. The reaction forms a floating slag, which can be skimmed off. Desulfurization may take place at various locations within the iron and steel-making facility; however, if the location is the BOF shop, then this process is most often accomplished at the hot-metal–transfer (reladling) station to take advantage of the fume collection system at that location.

GHG Emissions

The vast majority of GHGs (CO_2) is emitted from the blast furnaces' stove stacks where the combustion gases from the stoves are discharged. A small amount of emissions may also occur from flares, leaks in the ductwork for conveying the gas, and from blast furnace emergency venting. Emissions of CO_2 are also generated from the combustion of natural gas using flame suppression to reduce emissions of PM. In flame suppression, a flame is maintained over the surface of the molten metal, for example, during tapping, to consume oxygen and inhibit the formation of metal oxides that become airborne. Emissions also occur from the combustion of BFG in flares (flaring).

The IPCC Guidelines provide an emissions factor of 260 tonnes of CO_2 per terajoule for the combustion of BFG. Based on the production of 39.8 million tons (36.1 million tonnes) of pig iron in 2007, CO_2 emissions from blast furnace stoves would be approximately 26 million tons (24 million tonnes) of CO_2/yr.

A.3.3. Steelmaking Process—Basic Oxygen Furnaces (EPA, 2001b; 2008b)

The BOF is a large, open-mouthed pear-shaped vessel lined with a basic refractory material that refines iron into steel. The term "basic" refers to the chemical characteristic or pH of the lining. The BOF receives a charge composed of molten iron from the blast furnace and ferrous scrap. The charge is typically 70 percent molten iron and 30 percent steel scrap. A jet of high-purity oxygen is injected into the BOF, which oxidizes the carbon and silicon in the molten iron to remove these constituents and to provide heat for melting the scrap. After the oxygen jet is started, lime is added to the top of the bath to provide a slag of the desired pH or basicity. Fluorspar (a mineral) and "mill scale" (an iron oxide waste material generated by rolling mills) are also added to achieve the desired slag fluidity. The oxygen combines with the unwanted elements (with the exception of sulfur) to form oxides, which leave the bath as gases or enter the slag. As refining continues and the carbon content decreases, the melting point of the bath increases. Sufficient heat must be generated from the oxidation reactions to keep the bath molten.

The distinct operations in the BOF process are the following:

- Charging—Adding molten iron and metal scrap to the furnace;
- Oxygen blow—Introducing oxygen into the furnace to refine the iron;
- Turndown—Tilting the vessel to obtain a sample and check temperature;
- Reblow—Introducing additional oxygen, if needed;
- Tapping—Pouring the molten steel into a ladle; and
- Deslagging—Pouring residual slag out of the vessel.

There are currently three methods that are used to supply the oxidizing gas: (1) top blown, (2) bottom blown, and (3) combination blowing. Most bottom-blown furnaces use tuyeres consisting of two concentric pipes, in which oxygen is blown through the center of the inner pipe and a hydrocarbon coolant (such as CH_4) is injected between the two pipes. The hydrocarbon decomposes at the temperature of liquid steel, absorbing heat as it exits and protecting the oxygen tuyere from overheating and burn back.

In the BOF process, molten iron from a blast furnace and iron scrap are refined in a furnace by lancing (or injecting) high-purity oxygen. In this thermochemical process, careful computations are made to determine the necessary percentage of molten iron, scrap, flux materials, and alloy additions. Various steel-making fluxes are added during the refining process to reduce the sulfur and phosphorus content of the metal to the prescribed level. The oxidation of silicon, carbon, manganese, phosphorus, and iron provide the energy required to melt the scrap, form the slag, and raise the temperature of the bath to the desired temperature. The oxygen reacts with carbon and other impurities to remove them from the metal. Because the reactions are exothermic, no external heat source is necessary to melt the scrap and to raise the temperature of the metal to the desired range for tapping. The large quantities of CO produced by the reactions in the BOF can be controlled by combustion at the mouth of the furnace and then vented to gas-cleaning devices, as with open hoods, or combustion can be suppressed at the furnace mouth, as with closed hoods. The full furnace cycle typically takes 25 to 45 minutes.

The major emission point for CO_2 from the BOF is the furnace exhaust gas that is discharged through a stack after gas cleaning. The carbon is removed as CO and CO_2 during the oxygen blow. Carbon may also be introduced to a much smaller extent from fluxing materials and other process additives that are charged to the furnace. Using the default values in the IPCC Guidelines for iron (0.04) and steel (0.01) for the fraction of carbon gives an emission factor of 0.11 ton of CO_2/ton of steel (0.11 tonne of CO_2/tonne of steel) for carbon removed from the iron as CO_2. Applying the emission factor to the production of 44 million tons (40 million tonnes) of steel in BOFs in 2007 yields an estimate of 4.9 million tons (4.4 million tonnes) of CO_2/yr.

A.3.4. Steelmaking Process—Electric Arc Furnace (EPA, 2008b)

Electric arc furnaces are used to produce carbon and alloy steels. These steel-making furnaces are operated as a batch process that includes charging scrap and other raw materials, melting, removing slag ("slagging"), and tapping. The length of the operating cycle is referred to as the tap-to-tap time, and each batch of steel produced is known as a "heat." Tap-to-tap times range from 35 minutes to more than 200 minutes, with generally higher tap-to-tap times for stainless and specialty steel. Newer EAFs are designed to achieve a tap-to-tap time of less than 60 minutes.

The input material to an EAF is typically 100 percent scrap. Cylindrical refractory-lined EAFs are equipped with carbon electrodes to be raised or lowered through the furnace roof. With electrodes retracted, the furnace roof can be rotated aside to permit the charge of scrap steel by overhead crane. After ferrous scrap is charged to the EAF, the melting phase begins when electrical energy is supplied to the carbon electrodes. Electric current of the opposite polarity electrodes generates heat between the electrodes and through the scrap. Oxy-fuel burners and oxygen lances may also be used to supply chemical energy. Oxy-fuel burners, which burn natural gas and oxygen, use convection and flame radiation to transfer heat to the scrap metal. Oxygen lances are used to inject oxygen directly into the molten steel; exothermic reactions with the iron and other components provide additional energy to assist in melting the scrap and removing excess carbon. Alloying agents and fluxing materials usually are added through the doors on the side of the furnace to achieve the desired composition.

Refining of the molten steel can occur simultaneously with melting, especially in EAF operations where oxygen is introduced throughout the batch. During the refining process, substances that are incompatible with iron and steel are separated out by forming a layer of slag on top of the molten metal. After completion of the melting and refining steps, the slag door is opened, and the furnace is tipped backward so the slag pours out ("slagging"). The furnace is righted, and the tap hole is opened. The furnace is then tipped forward, and the steel is poured ("tapped") into a ladle (a refractory-lined vessel designed to hold the molten steel) for transfer to the ladle metallurgy station. Bulk alloy additions are made during or after tapping based on the desired steel grade.

Some EAF plants, primarily the small specialty and stainless steel producers, use AOD to further refine the molten steel from the EAF to produce low-carbon steel. In the AOD vessel, argon and oxygen are blown into the bottom of the vessel, and the carbon and oxygen react to form CO_2 and CO, which are removed from the vessel.

The CO_2 emissions are generated during the melting and refining process when carbon is removed from the charge material and carbon electrodes as CO and CO_2. These emissions are captured and sent to a baghouse for removal of PM before discharge into the atmosphere. The AOD vessels are small contributors to CO_2 emissions.

The CO_2 emissions estimate of 5.1 million tons (4.6 million tonnes) of CO_2 for EAFs is based on the IPCC Guidelines emission factor of 0.08 ton of CO_2/ton of steel (0.08 tonne of CO_2/tonne of steel) and the production of 64 million tons (58 million tonnes) of steel in 2007.

A.3.5. Casting and Finishing (EPA, 2008b) Casting

The steel produced by both BOFs and EAFs follow similar routes after the molten steel is poured from the furnace. The molten steel is transferred from ladle metallurgy to the continuous caster, which casts the steel into semi-finished shapes (e.g., slabs, blooms, billets, rounds, other special sections). Continuous casting is a relatively recent development, which has essentially replaced the ingot casting method because it increases the process yield from 80 percent to more than 95 percent and offers significant product quality benefits. Continuous casting has also decreased GHG emissions due to the increased yield and from a decrease in energy use as compared to energy-intensive ingot casting. Continuous casting is used to produce approximately 99 percent of the steel today. Both continuous and ingot casting are not estimated to be significant sources of GHGs.

Ingot casting was the common casting route prior to continuous casting, and only a small amount of steel is now processed using this route. In this process, molten steel is poured from the ladle into an ingot mold, where it cools and begins to solidify. The molds are stripped away, and the ingots are transported to a soaking pit or to a reheat furnace where they are heated to a uniform temperature. The ingots are shaped by rolling them into semi-finished products, usually slabs, blooms, or billets, or by forging. Ingot casting is typically used for small specialty batches and certain applications for producing steel plates.

Whichever production technique is used, the slabs, blooms, or billets undergo a surface preparation step, called "scarfing," which removes surface defects before shaping or rolling. Scarfing can be performed by a machine applying jets of oxygen to the surface of hot semi-finished steel or by hand (with torches) on cold or slightly heated semi-finished steel.

Rolling Mills

Steel from the continuous caster is processed in rolling mills to produce steel shapes that are classified according to general appearance, overall size, dimensional proportions, and intended use. Slabs are always oblong, usually 2- to 9-in. thick and 24- to 60-in. wide (5- to 23- centimeter [cm] thick and 61- to 152-cm wide). Blooms are square or slightly oblong and are mostly in the range of 6-by-6 in. to 12-by-12 in (15-by-15 cm to 30-by-30 cm). Billets are mostly square and range from 2-by-2 in. to 5-by-5 in (5-by-5 cm to 13-by-13 cm). Rolling mills are used to produce the final steel shapes that are sold by the steel mill. These shapes include coiled strips, rails, and other structural shapes, as well as sheets and bars. Because rolling mills consume electricity, they consequently contribute to indirect emissions of GHGs.

A.3.6. Other Steel Finishing Processes and Combustion Sources (EPA, 2008b)

The semi-finished products may be further processed by using many different steps, such as annealing, hot forming, cold rolling, pickling, galvanizing, coating, or painting. Some of these steps require additional heating or reheating. The additional heating or reheating is accomplished using furnaces usually fired with natural gas. The furnaces are custom designed for the type of steel, the dimensions of the semi-finished steel pieces, and the desired temperature.

There are many different types of combustion processes at both integrated iron and steel and EAF steel facilities that are not directly related to the major production processes previously discussed. However, the EAF facilities burn natural gas almost exclusively, whereas integrated facilities burn a combination of fuels, including natural gas, COG, and BFG. The combustion units at both types of facilities include boilers, process heaters, flares, dryout heaters, and several types of furnaces. For example, soaking pits and reheat furnaces are used to raise the temperature of the steel until it is sufficiently hot to be plastic enough for economical reduction by rolling or forging. Annealing furnaces are used to heat the steel to relieve cooling stresses induced by cold or hot working and to soften the steel to improve machinability and formability. Ladle reheating uses natural gas to keep the ladle hot while waiting for molten steel. Natural gas is the most commonly used fuel, in general, at both types of steel-making facilities, but COG and BFG (depending on availability) are also used in some of the combustion processes at integrated plants. The CO_2 emissions from combustion sources in 2007 were estimated at approximately 21 million tons (19 million tonnes) for EAF steel plants and 19 million tons (17 million tonnes) for integrated iron and steel plants.

A.3.7. Coke Production (EPA, 2008b, 2001a)

Most coke is produced in by-product recovery coke oven batteries. However, of the 19 U.S. coke plants shown in Table 1, there are four non-recovery coke oven batteries, including the three newest coke plants. All three of the newest non-recovery plants use waste heat from combustion to generate electricity. The recovery of waste heat to generate electricity reduces the amount of purchased electricity or reduces the need to purchase additional fuel to generate electricity onsite. Recovered heat that is supplied to the grid also reduces the amount of electricity that must be produced; if this power is generated from fossil-fuel combustion, then the recovered heat lowers the amount of CO_2 emissions generated from combustion.

By-product Recovery Coke Oven Batteries

Thermal distillation is used to remove volatile non-carbon elements from coal to produce coke in ovens grouped together in "batteries." A by-product coke oven battery consists of 20 to 100 adjacent ovens with common side walls made of high-quality silica and other types of refractory brick. The wall separating adjacent ovens and each end wall consists of a series of heating flues. At any one time, half of the flues in a given wall will be burning gas in combustion flues, and the other half of the flues will be conveying waste heat from the combustion flues to a heat exchanger and then to the combustion stack. Every 20 to 30 minutes, the battery "reverses," the former waste heat flues become combustion flues, and the former combustion flues become waste heat flues. Because the flame temperature is above the melting point of the brick, this reversal avoids melting the battery brickwork and provides more uniform heating of the coal mass. Process heat is obtained from the combustion of COG in the combustion flues, which is sometimes supplemented with BFG. The BFG is introduced from piping in the basement of the battery where the gas flow to each flue is metered and controlled. Waste gases from combustion, including GHGs, exit through the battery stack.

Each oven holds between 15 and 25 tons (14 and 23 tonnes) of coal. Offtake flues remove gases evolved from the destructive distillation process. The operation of each oven in the battery is cyclic, but the batteries usually contain a sufficiently large number of ovens so that the yield of by-products is essentially continuous. Coking continues for 15 to 18 hrs to produce blast furnace coke and 25 to 30 hrs to produce foundry coke. The coking time is determined by the coal mixture, the moisture content, the rate of underfiring, and the desired properties of the coke. Coking temperatures generally range from 1,700°F to 2,000°F (900°C to 1,100°C) and are kept on the higher side of the range to produce blast furnace coke.

The coke oven process begins with pulverized coal that is mixed and blended, with water and oil sometimes added to control the bulk density of the mixture. The prepared coal mixture is then transported to the coal storage bunkers on the coke oven battery. A specific volume of coal is discharged from the bunker into a larry car, which is a vehicle that moves along the top of the battery. When the larry car is positioned over an empty, hot oven, the lids on the charging ports are removed, and the coal is discharged from the hoppers of the larry car into the oven. To minimize the escape of gases from the oven during charging, steam aspiration is used to draw gases from the space above the charged coal into a collecting main duct. After charging, the aspiration is turned off, and the gases are directed through an offtake system into the gas-collecting main duct.

The maximum temperature attained at the center of the coke mass usually ranges from 2,000°F to 2,800°F (1,100°C to 1,500°C). At this temperature, almost all volatile matter from the coal mass volatilizes and leaves a high-quality metallurgical coke. Ambient air is prevented from leaking into the ovens by maintaining a slight positive back pressure of approximately 10 mm of water. The positive pressure causes some COG to leak out of the ovens. The gases and hydrocarbons, including GHGs, that evolve during thermal distillation in the coke oven are removed through the offtake gas system and are sent to the by-product plant for recovery.

Near the end of the coking cycle, each oven is disconnected, or "dampered off," from the main collection duct. Once an oven is dampered off, a standpipe in the oven that is capped during the cycle is opened to relieve pressure. Volatile gases exit through the open standpipe and are ignited if they fail to self-ignite. These gases are allowed to burn until the oven has been emptied of coke, or "pushed." At the end of the coking cycle, doors at both ends of the oven are removed, and the hot coke is pushed out of the coke side of the oven by a ram that is extended from a pusher machine. The coke is then pushed through a guide trough into a special rail car (called a quench car), which traverses the coke side of the battery. The quench car carries the coke to a quench tower where the hot coke is deluged with water. The quenched coke is discharged onto an inclined "coke wharf" to allow excess water to drain and cool the coke to a lower temperature. Gates along the lower edge of the wharf control the rate that the coke falls onto a conveyor belt that carries it to a crushing and screening system.

Gases that evolve during coking leave the coke oven through standpipes, pass into goosenecks (curved piping that connects each oven's standpipe to the main collecting duct), and travel through a damper valve to the gas collection main duct that directs the gases to the byproduct plant. These gases account for 20 to 35 percent by weight of the initial coal charge and are composed of water vapor, tar, light oils, heavy hydrocarbons, and other chemical compounds.

At the by-product recovery plant, tar and tar derivatives, ammonia, and light oil are extracted from the raw COG. At most coke plants, after tar, ammonia, and light oil are removed, the gas undergoes a final desulfurization process to remove hydrogen sulfide before being used as fuel. Approximately 35 to 40 percent of cleaned COG (after the removal of economically valuable by-products) is used to heat the coke ovens, and the remainder is used in other operations related to steel production, in boilers, or is flared. COG is composed of approximately 47 percent hydrogen, 32 percent CH_4, 6 percent CO, and 2 percent CO_2.

Non-Recovery Coke Oven Batteries

As the name implies, the non-recovery cokemaking process does not recover the numerous chemical by-products, as previously discussed. All of the COG is burned, and instead of recovering the chemicals, this process offers the potential for heat recovery and cogeneration of electricity. Nonrecovery ovens are of a horizontal design (as opposed to the vertical slot oven used in the by-product process) with a typical range of 30 to 60 ovens per battery. The oven is generally between 30- and 45-feet (ft.) (9 and 14-meters [m]) long and 6- to 12-ft. (1.8- to 3.7-m) wide. The internal oven chamber is usually semi-cylindrical, with the apex of the arch 5 to 12 ft. (1.5 to 3.7 m) above the oven floor. Each oven is equipped with two doors, one on each side of the horizontal oven, but there are no lids or offtakes as found on by-product ovens. The oven is charged through the oven doorway with a coal conveyor rather than from the top through charging ports as in a recovery plant.

After a non-recovery oven is charged with coal, carbonization begins as a result of the heat radiated from the oven bricks used with the previous charge. Combustion products and volatiles that evolve from the coal mass are burned in the chamber above the coal, in the gas pathway through the walls, and beneath the oven in combustion flues ("sole" flues). Each oven chamber has two to six "downcomers" ducts in each oven wall; the sole flue may be subdivided into separate flues that are supplied by these downcomers. The sole flue is designed to heat the bottom of the coal charge by conduction, and radiant and convective heat flow is produced above the coal charge.

Primary combustion air is introduced into the oven chamber above the coal (the "crown") through one of several dampered ports in the door. The dampers are adjusted to maintain the proper temperature in the oven crown. Outside air may also be introduced into the sole flues; however, additional air is usually required in the sole flue only for the first hour or two after charging. All of the non-recovery ovens are maintained under a negative pressure and do not leak under normal operating conditions, unlike the by-product ovens, which are maintained under a positive pressure. The combustion gases are removed from the ovens and directed to the stack through a waste heat tunnel located on top of the battery centerline and extends the length of the battery.

GHG Emissions from Coke Plants

The primary emissions point of gases at coke plants is the battery's combustion stack. Test data were obtained for 53 emissions tests (generally three runs per tests) for CO_2 emissions from the combustion stacks at by-product recovery coke plants for development of an emissions factor for EPA's 2008 revision to AP-42 (EPA, 2008a). These tests averaged 0.21 ton of CO_2/ton of coke (0.21 tonne of CO_2/tonne of coke). Test results for a non-recovery battery were also obtained and analyzed. The average of three runs at Haverhill Coke resulted in an emissions factor of 1.23 ton of CO_2/ton of coke (1.23 tonne of CO_2/tonne of coke), approximately six times higher than the factor for the combustion stack at by-product recovery batteries. The emissions factor for non-recovery combustion stacks is much higher because all of the COG and all of the by-products are burned. In comparison, organic liquids (e.g., tar, light oil) are recovered at byproduct recovery coke plants, and only approximately one-third of the gas is consumed in underfiring the ovens. Emissions from combustion stacks based on the 2007 production rate are estimated at 3.3 million tons (3 million tonnes) of CO_2 from non-recovery battery stacks at three coke plants and 3.1 million tons (2.8 million tonnes) of CO_2 from by-product recovery battery stacks at the 15 U.S. coke plants.

A small amount of CO_2 is emitted from the pushing operation when the incandescent coke is pushed from the oven and transported to the quench tower where it is quenched with water. The 2008 revisions to EPA's AP-42 compilation of emissions factors provide an emissions factor of 0.008 ton of CO_2/ton of coal (0.008 tonne of CO_2/tonne of coal), which is equivalent to 0.01 ton of CO_2/ton of coke (0.01 tonne of CO_2/tonne of coke) (EPA, 2008a). Using the 2007 production rate for coke 17.4 million tons (15.8 million tonnes), the emissions from pushing are estimated at 0.174 million tons (0.158 million tonnes) of CO_2/yr.

Fugitive emissions occur during the coking process from leaks of raw COG that contains CH_4. The leaks occur from doors, lids, offtakes, and collecting mains and are almost impossible to quantify because they change in location, frequency, and duration during the coking cycle, and they are not captured in a conveyance. However, the number, size, and frequency of these leaks have decreased significantly over the past 20 years as a result of stringent regulations, including national standards, consent decrees, and state regulations.

Many by-product recovery coke plants also have other combustion sources, primarily boilers and flares. These units use excess COG that is not used for underfiring the battery or shipped offsite for use as fuel in other processes. The IPCC Guidelines provide an emissions factor of 0.56 ton of CO_2/ton of coke (0.56 tonne of CO_2/tonne of coke) (assuming all of the COG is burned). Emissions from the combustion of COG in units other than the coke battery underfiring system are estimated at 0.35 ton of CO_2/ton of coke (0.35 tonne of CO_2/tonne of coke). For the production of 8.7 million tons (7.6 million tonnes) of coke in stand-alone by-product coke plants (i.e., the nine by-product coke plants not located at integrated iron and steel facilities), emissions from other combustion units would be 3.0 million tons (2.7 million tonnes) of CO_2/yr.

A.4. Energy Intensity

As shown in Figure A-1, the U.S. iron and steel industry has reduced overall energy intensity for steel production dramatically since the 1950s (AISI, 2010). A large part of this decrease is due to the increasing proportion of steel recycled in EAFs since the 1970s—the energy intensity of secondary steelmaking is much less than the integrated iron and steel route (19 MMBtu/ton [22 GJ/tonne] for integrated and 5.0 MMBtu/ton [6 GJ/tonne] for EAFs). Some of the other contributors to the reduction include the widespread adoption of continuous casting, blast furnace coal injection, optimization of blast furnace operations, thin-slab casting, and the use of previously wasted process gases (BFG and COG) in furnaces and boilers.

As shown in Table A-5, several plants have installed cogeneration systems (EIA/DOE, 2003). The three newest coke plants all recover the heat from the battery stack to produce steam and/or electricity. Integrated iron and steel plants use excess process fuel gases (BFG and COG) for cogeneration units.

Many plants have implemented thin-slab casting (see Table A-6), where thin slabs are slabs that are 2- to 4-in. (5- to 10-cm) thick. This technology may be a future opportunity for a few more plants that produce thin slabs of steel. Thin-slab casting integrates casting and hot rolling into one process, which is estimated to reduce energy consumption by 4.2 MMBtu/ton (4.9 GJ/tonne) of crude steel.

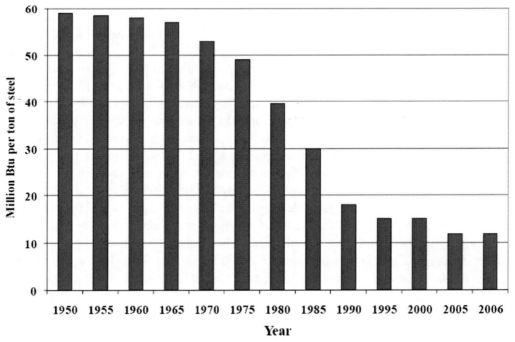

Source: American Iron and Steel Institute.

Figure A-1. Historical energy consumption in the iron and steel industry.

Table A-5. Electricity Generation in the Iron and Steel Industry

Plant	Nameplate Capacity
Coke Plants	
Indiana Harbor Coke, East Chicago, IN[a]	94
Haverhill Coke, Haverhill, OH[b]	46
US Steel, Clairton, PA[c]	81
Gateway Energy and Coke, Granite City, IL[d]	139
Erie Coke, Erie, PA[c]	2.5
Total	362.5
Steel Plants	
Arcelor Mittal (was Ispat), East Chicago, IN[c]	263
Arcelor Mittal (was LTV), East Chicago, IN[c]	97
Arcelor Mittal (was Bethlehem), Burns Harbor, IN[c]	178
Arcelor Mittal (was LTV), Cleveland, OH[c]	45
Severstal (was WCI), Warren, OH[c]	20.5
Severstal (was Bethlehem), Sparrows Point, MD[c]	170
US Steel, Fairfield, AL[c]	82
US Steel, Gary, IN[c]	231
Total	1,086.5

a Compiled from press releases. The cogeneration plant is owned by Primary Energy, a subsidiary of NIPSCO.

b Compiled from press releases. The Phase 1 coke batteries recovered the heat to produce steam for a nearby Sunoco chemical plant. The cogeneration unit was installed for the Phase 2 batteries to produce electricity for sale.

c From EIA/DOE, *Inventory of Nonutility Electric Power Plants in the United States 2000.* Energy Information Administration, U.S. Department of Energy. DOE/EIA-0095(2000)/2. January 2003.

d Compiled from press releases. Planned to be fully operational in 2010.

Table A-6. U.S. Slab Casting Units Installed 1994–2000[a]
with Thin Slab Facilities Noted

Facility	City	State	No. of Units	Year of Startup	Annual Capacity (1,000 tpy)	Product Thickness Range (in.)	Median Product Thickness (in.)
Mini-mills							
Nucor Steel[b]	Crawfordsville	IN	1	1994	1,000	2	2.0
Nucor Steel[b]	Hickman	AR	1	1994	1,000	2	2.0
Gallatin Steel[b]	Ghent	KY	1	1995	1,450	2.2–2.6	2.4
AK Steel[b]	Mansfield	OH	1	1995	750	3–5	4.0
Corus Tuscaloosa	Tuscaloosa	AL	1	1996	880	5.1	5.1

Table A-6. (Continued)

Facility	City	State	No. of Units	Year of Startup	Annual Capacity (1,000 tpy)	Product Thickness Range (in.)	Median Product Thickness (in.)
Steel Dynamics Inc.[b]	Butler	IN	1	1996	2,400	2.16	2.2
North Star BHP Steel	Delta	OH	1	1996	1,560	6.5	6.5
Beta Steel Corp.[b]	Portage	IN	1	1997	700	3.5	3.5
Ipsco Steel	Muscatine	IA	1	1997	1,250	5–6	5.5
Nucor Steel[b]	Berkeley Cnty	SC	1	1997	2,700	2.1–2.4	2.2
Steel Dynamics Inc.[b]	Butler	IN	1	1998	2,400	2.16	2.2
Nucor Steel[b]	Berkeley Cnty	SC	1	2000	2,700	2.1–2.4	2.2
Integrated Mills							
Geneva Steel	Provo	UT	1	1994	2,500	8.6	8.6
Rouge Steel Co.	Dearborn	MI	1	1996	1,300	8.0	8.0
Ispat Inland Ind Hbr	East Chicago	IN	1	2000	3,000	9.25	9.25
International Steel Gr	Sparrows Pt	MD	1	2000	2,200	10–12	11.0

[a] Data taken from the 2003 Continuous Caster Roundup (AIST, 2003).
[b] Uses thin-slab casting.

APPENDIX B. EMERGING TECHNOLOGIES FROM U.S. DEPARTMENT OF ENERGY (DOE) AND DOE PARTNERSHIPS

I. Active Research and Development Projects (DOE, 2009a)

The following projects are currently underway to improve the energy efficiency, environmental performance, and productivity of the steel industry. Emerging technologies are defined as technologies that are likely to be commercially available in the next 2 years.

Advanced Process Development
- Minimization of Blast Furnace Fuel Rate by Optimizing Burden and Gas Distributions
  - **Partners:** Purdue University Calumet, AISI, Mittal Steel, Dofasco, and Severstal.
  - **Summary**: A computational fluid dynamics (CFD) model will help to optimize and burden distributions that can minimize fuel rate, thereby maximizing blast furnace energy efficiency and minimizing emissions.

- o **Benefits:** Increase pulverized coal injection rate and fuel efficiency, reduces carbon emissions, and optimizes blast furnace efficiency.
- o **Status (August 2007):** The project team has conducted an initial market study and developed a marketing plan. There are 28 blast furnaces currently operating in the U.S., of which, 13 are operated by this project's industrial partners. The newly developed CFD technology will be implemented in each industrial partner's blast furnace during the project period. Within 5 years of successful project completion, the remaining blast furnaces in the U.S. will be targeted for implementation. A final marketing and technology transfer plan will be developed as part of the final deliverables of this project.
- Research, Development, and Field Testing of Thermochemical Recuperation for High-Temperature Furnaces
  - o **Status:** The contract ($4.5 million) was awarded in September 2008. AISI is leading a team with the Gas Technology Institute, Thermal Transfer Corporation, U.S. Steel, ArcelorMittal, Republic Engineered Products, the Steel Manufacturing Association, and the Ohio Department of Development to develop and test thermochemical recuperation for steel reheating furnaces to increase waste heat recovery that reduces energy consumption and costs. A thermochemical recuperator uses the partialoxidation-of-fuel principle to recover energy from flue gases of heating processes.

Cokeless Ironmaking
- Next Generation Metallic Iron Nodule Technology in Electric Furnace Steelmaking – Partners: University of Minnesota-Duluth and Nu-Iron Technologies, LLC.
  - o **Benefits:** Metallic iron nodule technology produces a high -quality scrap substitute, reduces production costs, increases steel quality produced by EAFs, and enables more effective use of sub-bituminous coal. Successful development of this new ironmaking process will produce potentially lower cost steel scrap substitutes. Also, greater availability of high-quality iron nodules will increase the quality of steel and the competitiveness of mini-mills and other steel producers.
  - o **Status (September 2007):** Phase 1 is complete. The testing phase will involve quantifying overall energy use characteristics, types of material that can be processed, fuels needed for successful operation, and the overall economics predicted for full-scale implementation. Upon successful demonstration, the project team will begin plans to transition the technology for industrial use. Iron nodule technology could potentially use up to 30 percent less energy than utilizing rotary hearth furnace technology.
- Paired Straight Hearth Furnace
  - o – **Status:** The Phase 1 report (feasibility study) was completed in February 2006. The Bricmont, Inc., report and the McMaster University analysis concluded that it is feasible with current technology and construction practices to design, build, and operate a demonstration plant of the PSH furnace with a capacity of 46,000 ton per year (42,000 tonne per year) of DRI for an estimated cost of $16,729,000. A DOE contract ($1.5 million) was awarded in September 2008. AISI, in partnership with McMaster University, U.S. Steel, Bricmont, and

Harper International, will work to optimize the PSH furnace technology and establish its scalability potential from the bench-scale stage. The PSH furnace is an alternative to the energy and carbon-intensive blast furnace commonly used to make steel. The technology has a lower coal rate in comparison with other

Status: The Phase 1 report (feasibility study) was completed in February 2006. The Bricmont, Inc., report and the McMaster University analysis concluded that it is feasible with current technology and construction practices to design, build, and operate a demonstration plant of the PSH furnace with a capacity of 46,000 ton per year (42,000 tonne per year) of DRI for an estimated cost of $16,729,000. A DOE contract ($1.5 million) was awarded in September 2008. AISI, in partnership with McMaster University, U.S. Steel, Bricmont, and Harper International, will work to optimize the PSH furnace technology and establish its scalability potential from the bench-scale stage. The PSH furnace is an alternative to the energy and carbon-intensive blast furnace commonly used to make steel. The technology has a lower coal rate in comparison with other alternative ironmaking processes because of thermodynamic and kinetic advantages.

Next Generation Steelmaking

- Development of Next Generation Heating System for Scale-Free Steel Reheating, Phase 2
  - o **Partners:** E3M, Inc.; ACL-NWO, Inc.; Bloom Engineering Corp.; Steel Dynamics, Inc.; Air Products & Chemical; the Steel Manufacturers Association; and the Forging Industry Association.
  - o **Benefits:** Scale-free reheating improves productivity by reducing downtime and manpower to collect and remove scale. Scale-free reheating increases energy and cost efficiency of steel reheating and reduces the amount of energy needed to replenish steel lost as oxides. By reducing the amount of steel lost to scale formation, this system improves the surface quality of the steel.
  - o **Status (September 2007):** Completed Phase 1, which included three activities: (1) conducting a literature search and analyzed the options needed to create a process atmosphere required for scale-free reheating, (2) defining furnace operating parameters required to generate scale-free heating process atmosphere, and (3) conducting economic and technical analyses. Phase 2 will include conducting pilot-scale furnace heating tests on scale-free heating, defining heating system conditions, designing and validating a scale-free heating system for typical applications, and conducting energy, economic, and environmental analyses and modeling. During the commercialization phase, the scale-free heating burner will be tested for functionality in furnaces used for both conventional heating and scale-free heating.

II. Success Stories (DOE, 2009d)

Collaborative R&D projects under the auspices of the Steel Industry of the Future have produced energy, environmental, and economical benefits for the industry and the nation. The following list contains examples of projects that have been commercially successful and demonstrated full-scale or completed industrial trials:

- Enhanced Spheroidized Annealing
- Mesabi Nugget Ironmaking Technology for the Future: High Quality Iron Nuggets Using a Rotary Hearth Furnace
- Dilute Oxygen Combustion
- Hot-Blast Stove Process Model
- Microstructure Engineering in Hot Strip Mills
- Nickel Aluminide Transfer Rolls
- NO_x Emission Reduction by Oscillating Combustion
- Development of a Process to Continuously Melt, Refine, and Cast High-Quality Steel

III. Completed Research and Development Projects (DOE, 2009c)

The following projects were recently completed. In some cases, the R&D produced a new technology that is now emerging in the marketplace. In other cases, the R&D results will help to guide future development of energy-efficient technologies and processes for the steel industry.

- Advanced Control in Blast Furnace
- Aluminum Bronze Alloys to Improve the System Life of Basic Oxygen and EAF Hoods, Roofs, and Side Vents
- Appropriate Resistance Spot Welding Practice for Advanced High-Strength Steels
- Automated Steel Cleanliness Analysis Tool
- CFD Modeling for High-Rate Pulverized Coal Injection in the Blast Furnace
- Characterization of Fatigue and Crash Performance of a New Generation of High Strength Steel
- Clean Steels: Advancing the State of the Art
- Cold Work Embrittlement of Interstitial-Free Steels
- Constitutive Behavior of High-Strength Multiphase Sheet Steels Under High-Strain Rate Deformation Conditions
- Controlled Thermal-Mechanical Processing of Tubes and Pipes
- Dephosphorization When Using DRI or Hot Briquetted Iron
- Development and Application of Steel Foam and Structures
- Development of a Process to Continuously Melt, Refine, and Cast High-Quality Steel
- Development of Next Generation Heating System for Scale-Free Steel Reheating, Phase 1
- Development of Oxygen-Enriched Furnace
- Elimination or Minimization of Oscillation Marks—A Path to Improved Cast Surface Quality
- Enhanced Inclusion Removal from Steel in the Tundish
- Enrichment of By-Product Materials from Steel Pickling Acid Regeneration Plants
- Feasibility Study for Recycling Use Automotive Oil Filters in a Blast Furnace (Final Report)

- Formability Characterization of a New Generation of High-Strength Steels
- Future Steelmaking Processes (December 2003)
- Geological Sequestration of Carbon Dioxide (CO2) by Hydrous Carbonate Formation with Reclaimed Slag
- Hydrogen and Nitrogen Control in Ladle and Casting Operations
- Improved Criteria for Acceptable Yield Point Elongation of Surface Critical Steels
- Inclusion Optimization for Next-Generation Steel Products
- In Situ, Real-Time Measurement of Melt Constituents
- Integrating Steel Production with Mineral Sequestration
- Intelligent Inductive Processing
- Large-Scale Evaluation of Nickel Aluminide Rolls in a Heat-Treat Furnace
- Laser Contouring System
- Life Improvement of Pot Hardware
- Magnetic Gate System for Molten Metal Flow Control
- The Mesabi Nugget Research Project New Ironmaking Technology of the Future: High-Quality Iron Nuggets Using a Rotary Hearth Furnace
- Minimizing NO_x Emissions from By-Product Fuels in Steelmaking
- New Process for Hot-Metal Production at Low Fuel Rate—Phase 1 Feasibility Study
- New Ultra-Low–Carbon Steels with Improved Bake Hard Oak Ridge National Laboratory
- Novel Low-NO_x Burners for Boilers in the Steel Industry
- Optical Sensor for Post-Combustion Control in EAF Steelmaking
- Optimization of Post-combustion
- Plant Line Trial Evaluation of Viable Non-Chromium Passivation Systems for Electrolytic Tinplate
- Properties of Galvanized and Galvannealed Advanced High-Strength Hot-Rolled Steels
- Pulverized Coal Injection
- Quantifying the Thermal Behavior of Slags
- Real-Time Melt Temperature Measurement in a Vacuum Degasser Using Optical Optometry
- Recycling of Waste Oxides
- Removal of Residual Elements in the Steel Ladle
- Standard Methodology for the Quantitative Measurement of Steel Phase Transformation Kinetics
- Strip Casting: Anticipating New Routes to Steel Sheet
- Study of Deformation Behavior of Lightweight Steel Structures
- Submerged Entry Nozzles that Resist Clogging
- Suspension Hydrogen Reduction of Iron Oxide Concentrate
- Sustainable Steelmaking Using Biomass and Waste Oxides
- Technical Feasibility Study of Steelmaking by Molten Oxide Electrolysis
- Temperature Measurement of Galvanneal Steel
- Validation of the Hot Strip Mill Model
- Verification of Steelmaking Slag Iron Content

REFERENCES

AISI (American Iron and Steel Institute). 2010. Available at
http://www.steel.org/AM/Template.cfm?Section=Home&section=Energy2&template=/C
M/Cont entDisplay.cfm&ContentFileID=8600 and
http://www.steel.org//AM/Template.cfm?Section=Home.

AIST (Association for Iron & Steel Technology). 2003. AIST Roundup Series. AIST 2003
Continuous Caster Roundup. Available at
http://steellibrary.com/PartDetails/tabid/56/partid/08008204508007708304904904805 10
45051/ Default.aspx.

AIST (Association for Iron & Steel Technology). 2009. AIST 2009 Directory of Iron and
Steel Plants. Available at http://www.aist.org/publications/directory.htm.

CCAP (Center for Clean Air Policy). 2010. Global Sectoral Study: Final Report. Available at
http://www.ccap.org/index.php?component=programs&id=26.

DOE (U.S. Department of Energy). 2005. Steel Industry Marginal Opportunity Study.
Available at http://www1.eere.energy.gov/industry/steel/pdfs/steelmarginal
opportunity.pdf.

DOE (U.S. Department of Energy). 2008. Steel Success Story: Novel Technology Yields
High-Quality Iron Nuggets. Available at http://www1.eere.energy.gov/industry/steel/
pdfs/mnp success.pdf.

DOE (U.S. Department of Energy). 2009a. Active R&D Projects. Available at
http://www1.eere.energy.gov/industry/steel/active rd.html.

DOE (U.S. Department of Energy). 2009b. An Assessment of the Commercial Availability of
Carbon Dioxide Capture and Storage Technologies as of June 2009. Washington, DC.
U.S. Department of Energy, Office of Scientific and Technical Information. June.
Available at http://www.pnl.gov/science/pdf/PNNL-18520 Status of CCS 062009.pdf.

DOE (U.S. Department of Energy). 2009c. Completed R&D Projects. Available at
http://www1.eere.energy.gov/industry/steel/completed rd.html.

DOE (U.S. Department of Energy). 2009d. Success Stories. Available at
http://www1.eere.energy.gov/industry/steel/success.html.

EIA/DOE (Energy Information Administration/U.S. Department of Energy). 2003. Inventory
of Nonutility Electric Power Plants in the United States 2000. DOE/EIA-0095(2000)/2.
Available at http://www.eia.doe.gov/electricity/ipp/html/ippv2tc2p5.html.

EPA (U.S. Environmental Protection Agency). 2001a. National Emission Standards for
Hazardous Air Pollutants (NESHAP) for Coke Ovens: Pushing, Quenching, and Battery
Stacks—Background Information for Proposed Standards. EPA-453/R-01-006. Available
at http://www.epa.gov/ttn/atw/coke2/coke2p bid.pdf.

EPA (U.S. Environmental Protection Agency). 2001b. National Emission Standards for
Hazardous Air Pollutants (NESHAP) for Integrated Iron and Steel Plants—Background
Information for Proposed Standards. EPA-453/R-01-005. Available at
http://www.epa.gov/ttn/atw/iisteel/irnstlbid.pdf

EPA (U.S. Environmental Protection Agency). 2007a. Energy Trends in Selected
Manufacturing Sectors: Opportunities and Challenges for Environmentally Preferable
Energy Outcomes. Final Report. U.S. EPA; Office of Policy, Economics, and Innovation;

Sector Strategies Division, Research Triangle, Park, NC. March. Available at http://www.epa.gov/ispd/energy/index.html#report.

EPA (U.S. Environmental Protection Agency). 2007b. EPA Combined Heat and Power Partnership (CHPP). Catalog of CHP Technologies. Available at http://www.epa.gov/chp/basic/catalog.html.

EPA (U.S. Environmental Protection Agency). 2008a. AP-42 Section 12.2: Coke Production. Available at http://www.epa.gov/ttn/chief/ap42/ch12/final/c12s02 may08.pdf.

EPA/OAR (U.S. Environmental Protection Agency/Office of Air and Radiation). 2008b. Technical Support Document for the Iron and Steel Sector: Proposed Rule for Mandatory Reporting of Greenhouse Gases. Available at http://www.epa.gov/climatechange/emissions/archived/downloads/tsd/TSD%20Iron%20and%20 Steel%20 EPA%209-8-08.pdf.

Essar Steel Minnesota LLC. 2010. Welcome to Essar Steel Minnesota LLC. 2010. Available at http://www.essarsteelmn.com.

ICF. 2010. CHP Installation Database. Maintained for U.S. DOE and Oak Ridge National Laboratory.

R&D Magazine. 2010. ArcelorMittal Receives the Association for Iron & Steel Technology (AIST) 2009 Energy Achievement Award. Available at http://www.rdmag.com/News/Feeds/2010/02/materials-linde-oxyfuel-technology-helps-arcelormittal-win-a/.

Stubbles, J. 2000. Energy Use in the U.S. Steel Industry: An Historical Perspective and Future Opportunities. U.S. Department of Energy, Office of Industrial Technologies, Washington, DC. September. Available at http://www1.eere.energy.gov/industry/steel/pdfs/steel energy use.pdf.

Tenova. 2010. EAF Integration into the Blast Furnace Route at Wheeling Pittsburgh. Available at http://212.31.252.187/main project.php?id=81&id company=3&id prodotto=9&provenienza.

Worrell, E., N. Martin, and L. Price. 1999. Energy Efficiency and Carbon Dioxide Emissions Reduction Opportunities in the U.S. Iron and Steel Sector. (Report No. LBNL-41724). Ernest Orlando Lawrence Berkeley National Laboratory, Berkeley, CA. July. Available at http://www.energystar.gov/ia/business/industry/41724.pdf.

Worrell, E., P. Linde, M. Neelis, E. Blomen, and E. Masanet. 2009. Energy Efficiency Improvement and Cost Saving Opportunities for the U.S. Iron and Steel Industry. Ernest Orlando Lawrence Berkeley National Laboratory, Berkeley, CA. December.

Worrell, E. Personal communication. 2010. September 16.

INDEX

F

G

H